CONTRACTOR LICENSE EXAM PREP:

INTRODUCTION:

Welcome to the first step of an exciting journey, one that will lead you to the achievement of a significant milestone - becoming a licensed contractor. This book is not just a guide; it's a companion, a mentor, and a friend that will walk with you every step of the way.

We understand that embarking on this journey may feel daunting. You might be filled with aspirations, yet also clouded by uncertainties. You might have experienced setbacks and failures, and perhaps even fear the challenges that lie ahead. But remember, every successful contractor started where you are right now. They, too, faced the same fears, the same uncertainties, and the same challenges. And they prevailed. So can you.

This book is here to validate your uncertainties, to offer understanding in times of failure, and to provide comfort in the face of fear. It's here to stand alongside you against the adversities that may come your way. It's here to fuel your aspirations and to help you believe that your dream is not only possible but also worth every ounce of effort you're about to put in.

In the pages that follow, you'll find a wealth of knowledge, practical advice, and invaluable resources designed to equip you with the skills and confidence you need to ace the contractor's license exam. But more than that, you'll find stories of triumph, lessons from failures, and insights that will inspire and motivate you to keep going, even when the going gets tough.

So, as you turn the page and begin this journey, remember that every step you take brings you one step closer to your dream. And with each challenge you overcome, you're not only becoming a better contractor, you're also becoming a stronger, more resilient individual.

Let's embark on this journey together. Your dream is within reach, and this book will help you grasp it. Let's get started.

The importance of a contractor's license cannot be overstated. It's not just a piece of paper or a legal requirement. It's a testament to your skills, your knowledge, and your professionalism. It's a symbol of trust for your clients, assuring them that you adhere to the highest standards of quality and safety. It's a key that opens doors to new opportunities and growth in your career.

The process of obtaining a contractor's license may seem complex and daunting. There are eligibility criteria to meet, applications to fill out, exams to pass, and fees to pay. But don't worry, this book is here to guide you through every step of the process.

We'll start by helping you understand the eligibility criteria and the application process. We'll then delve into the details of the exam, covering all the topics you need to know, from business and law to trade skills and safety. We'll provide you with practice questions to test your knowledge and tips to help you perform your best on the exam day. And finally, we'll guide you through the steps to take after passing the exam, from obtaining your license to maintaining it.

The journey to becoming a licensed contractor begins with understanding the eligibility criteria. These criteria can vary by state, but generally, they include being at least 18 years old, having a high school diploma or equivalent, and possessing a certain amount of experience in the construction industry. Some states may also require you to be a U.S. citizen or a legal resident.

Once you've confirmed that you meet the eligibility criteria, the next step is the application process. This typically involves filling out an application form where you'll provide details about your education, work experience, and other relevant information. You may also need to submit supporting documents, such as proof of your work experience.

Along with your application, you'll need to pay certain fees. These can include an application fee, an examination fee, and a licensing fee. The exact amounts can vary by state, so it's important to check the specific requirements in your area. Keep in mind that these fees are usually non-refundable, even if your application is not approved.

It's important to approach the application process with care and attention to detail. Any errors or omissions can delay the process or even lead to your application being rejected. So, take your time, double-check all your information, and make sure you understand all the requirements before you submit your application.

Remember, obtaining a contractor's license is an investment in your future. The fees and costs associated with the process are just part of that investment. With your license, you'll be able to take on bigger projects, attract more clients, and grow your business. So, while the process may seem complex and costly, the potential rewards make it well worth the effort.

Understanding the structure and requirements of the contractor's license exam is crucial for effective preparation and success. Let's delve into these aspects.

The contractor's license exam typically consists of two parts: the Business and Law exam, which covers legal and business management aspects of contracting, and the Trade exam, which tests your knowledge and skills in the specific trade you're applying for. Each part is usually a multiple-choice test, and the number of questions can vary by state and trade.

The scoring system for the contractor's license exam also varies by state. Some states use a scaled scoring system, where your raw score (the number of questions you answered correctly) is converted to a scale that ranges from 0 to 100. Other states may simply use your raw score as your final score.

The passing score is typically around 70-75%, but this can vary by state and by exam. Some states may require a higher passing score for certain trades. It's important to check the specific requirements in your state to know exactly what score you need to aim for.

Remember, the exam is designed to test your knowledge and skills in a comprehensive way. It's not just about memorizing facts; it's about understanding concepts, applying knowledge, and making informed decisions. So, as you prepare for the exam, focus on understanding the material, not just memorizing it. With the right preparation, you can approach the exam with confidence and achieve the score you need to obtain your contractor's license.

Business and Law Section:

Contract law forms the backbone of the construction industry, governing the relationships, transactions, and disputes between parties involved in a construction project. At its core, contract law is about agreement and obligation. It's about making promises and fulfilling them.

In the construction industry, contracts serve several crucial functions. They define the scope of work, specify the terms of payment, allocate risks, and establish procedures for changes and dispute resolution. They provide a roadmap for the project, setting out what needs to be done, by whom, when, and for how much.

One of the fundamental principles of contract law is mutual assent, which means that all parties must agree to the terms of the contract. This is often expressed through the concept of "offer and acceptance". One party makes an offer, and the other party accepts it. In a construction contract, the offer might be a bid submitted by a contractor, and the acceptance might be the owner's decision to award the contract to that contractor.

Another key principle is consideration, which refers to something of value that is exchanged between the parties. In a construction contract, the consideration might be the contractor's promise to perform the work, and the owner's promise to pay for it.

A valid contract also requires capacity, meaning that the parties must have the legal ability to enter into a contract. For example, they must be of legal age and mentally competent.

Lastly, the contract must be for a lawful purpose. A contract for illegal construction work would not be enforceable.

Breach of contract occurs when one party fails to fulfill their obligations under the contract. In the construction industry, this could involve failing to complete the work on time, not following the specified plans and specifications, or not paying the contractor as agreed.

When a breach occurs, the non-breaching party has several potential remedies. They might seek damages, which is a monetary compensation for the loss caused by the breach. They might seek specific performance, which is a court order requiring the breaching party to fulfill their obligations. Or they might seek to terminate the contract and recover any payments they've made.

However, not all breaches allow for termination. The breach must be material, meaning it significantly affects the value or purpose of the contract. For example, if a contractor installs the wrong type of light fixtures, that might not be considered a material breach, because it doesn't significantly affect the overall project.

In some cases, the breaching party might have defenses to enforcement. For example, they might argue that the contract is unenforceable because it's vague or ambiguous, or because it was entered into under duress.

Understanding these principles of contract law can help you navigate the complex world of construction contracts, avoid disputes, and protect your rights and interests. Whether you're reviewing a contract, negotiating terms, or resolving a dispute, a solid grasp of contract law is an invaluable tool.

A legally binding contract requires four essential elements: agreement, consideration, contractual capacity, and lawful object.

1. **Agreement:** This involves an offer from one party and acceptance by another. For instance, a contractor offers to build a house for a certain price, and the homeowner accepts this offer.
2. **Consideration:** This is the value that each party gives to the other. It could be money, services, or even a promise to do or not do something. In our example, the consideration from the contractor is the promise to build the house, and from the homeowner, it's the promise to pay the agreed price.
3. **Contractual Capacity:** Both parties must have the legal ability to enter into a contract. They must be of legal age and sound mind. A contract with a minor or a mentally incapacitated person would not be legally binding.
4. **Lawful Object:** The purpose of the contract must be legal. A contract to perform illegal construction work would not be enforceable.

In the construction industry, various types of contracts are used, each with its unique characteristics:

1. **Express Contracts:** These are contracts in which the terms are explicitly stated by the parties, either in writing or orally. For example, a written agreement between a contractor and a homeowner where the contractor agrees to remodel the kitchen for a specified price is an express contract.
2. **Implied Contracts:** These are contracts that are not explicitly stated but are inferred from the actions of the parties. For instance, if a plumber visits a house, and the

homeowner allows them to fix a leak without discussing payment, an implied contract is formed. The homeowner is expected to pay a reasonable price for the service.

3. **Unilateral Contracts:** In these contracts, one party makes a promise in exchange for an act by the other party. For example, a contractor might offer a bonus to the workers if they complete the project by a certain date. The workers are not obligated to finish by that date, but if they do, the contractor is obligated to pay the bonus.
4. **Bilateral Contracts:** These are the most common type of contract in the construction industry. Both parties make promises to each other. For example, a contractor promises to build a house, and the homeowner promises to pay a certain amount for it.

Understanding these elements and types of contracts is crucial in the construction industry, as it helps to ensure that agreements are legally enforceable and that parties understand their rights and obligations.

In the realm of construction, a breach of contract occurs when one party fails to fulfill their obligations as outlined in the contract. This could manifest in various ways, such as:

1. **Non-performance:** This is the most straightforward type of breach. If a contractor fails to complete a project or a homeowner doesn't make the agreed-upon payments, they've breached the contract.
2. **Inferior performance:** If the contractor completes the project, but the work doesn't meet the quality standards specified in the contract, this could be considered a breach.
3. **Late performance:** Timing is crucial in construction. If the contractor doesn't complete the project within the agreed-upon timeframe, it's a breach of contract, unless the delay is due to reasons beyond their control.
4. **Violation of terms:** If either party violates any other terms of the contract, it's a breach. For example, if the contract requires the contractor to use a specific type of material, and they use a different one, they've breached the contract.

When a breach of contract occurs, the non-breaching party has several potential remedies:

1. **Damages:** The most common remedy, damages involve the breaching party paying the non-breaching party a certain amount to compensate for the breach. This could cover the cost of hiring a new contractor, the difference in value between the promised and delivered work, or any other financial losses resulting from the breach.
2. **Specific Performance:** In some cases, the non-breaching party may ask the court to order the breaching party to fulfill their obligations. This is more common when the subject of the contract is unique or rare, such as in contracts involving custom-made items.
3. **Cancellation and Restitution:** The non-breaching party may choose to cancel the contract and sue for restitution. This would return the parties to the position they were in before the contract, with the breaching party refunding any money or returning any property they received under the contract.
4. **Reformation:** In some cases, the court may choose to rewrite the contract to correct any inequities. This is typically used when the parties had a fundamental misunderstanding or when one party took advantage of the other.

Understanding these breaches and remedies can help you navigate potential disputes and protect your rights. It's also crucial for the contractor's license exam, as you'll need to demonstrate your understanding of contract law and its application in the construction industry.

Defenses to enforcement in contract law refer to the legal reasons or arguments that a party can use to avoid their contractual obligations. These defenses can be invoked when a party is accused of breaching a contract. Here are some common defenses used in the construction industry:

1. **Fraud:** This occurs when one party intentionally deceives another to get them to enter into a contract. For example, if a contractor misrepresents their qualifications or the quality of materials they will use, and the homeowner enters into the contract based on these false representations, the homeowner can use fraud as a defense if accused of breach.
2. **Duress:** Duress involves one party coercing another into a contract against their will. If a contractor threatens a homeowner to sign a contract, the homeowner can later argue that they signed the contract under duress, making it unenforceable.
3. **Mistake:** A mistake can be mutual (where both parties are mistaken about a fundamental aspect of the contract) or unilateral (where one party is mistaken). For instance, if both parties mistakenly believe that a certain type of material is available for a project, and it turns out it's not, this could be a mutual mistake that makes the contract unenforceable.
4. **Impossibility or Impracticability:** If fulfilling the contract becomes impossible or extremely difficult due to unforeseen circumstances, this can be a defense. For example, if a contractor is hired to renovate a building, but the building is destroyed in a natural disaster, the contractor could argue that performance is impossible.
5. **Unconscionability:** This refers to contracts that are so unfair to one party that they are oppressive. If a contract heavily favors the contractor and puts the homeowner at a severe disadvantage, the homeowner might be able to use unconscionability as a defense.
6. **Illegality:** If the contract involves illegal activities, it's unenforceable. For example, a contract for a construction project that violates building codes would be illegal.

Understanding these defenses is crucial for contractors. Not only can they help you protect your rights in a dispute, but they can also guide you in creating fair and enforceable contracts.

Written contracts are the lifeblood of the construction industry. They provide a clear record of what has been agreed upon, reducing the risk of misunderstandings and disputes. They detail the scope of work, payment terms, project timeline, and responsibilities of each party. They also provide a legal framework for enforcing these terms and resolving any disputes that may arise.

The process of contract negotiation and execution in a construction project involves several steps. First, the parties discuss the terms of the contract, such as the scope of work, price, and timeline. These discussions can involve a lot of back-and-forth as each party tries to secure the most favorable terms. Once the terms are agreed upon, they are put into a written contract.

The contract is then reviewed by both parties (and often their legal counsel) to ensure it accurately reflects the agreed-upon terms and complies with relevant laws and regulations. Any issues or concerns are addressed at this stage. Once both parties are satisfied with the contract, they sign it, making it legally binding.

Contract termination refers to the end of a contract. It can occur in several ways. First, a contract can be completed when both parties fulfill their obligations. For example, the contractor completes the work, and the homeowner makes the final payment.

Second, a contract can be terminated by mutual agreement. If both parties agree that they no longer wish to continue with the contract, they can agree to terminate it.

Third, a contract can be terminated due to a breach. If one party fails to fulfill their obligations, the other party may have the right to terminate the contract.

Finally, a contract can be terminated due to impossibility. If unforeseen circumstances make it impossible to fulfill the contract, it can be terminated.

The implications of contract termination depend on the reason for termination. If a contract is completed or terminated by mutual agreement, there are usually no negative implications. However, if a contract is terminated due to a breach or impossibility, it can lead to legal disputes and financial losses.

Understanding these aspects of contracts is crucial for anyone in the construction industry. They not only help you protect your rights and interests but also ensure that your projects run smoothly and successfully.

Let's consider a hypothetical scenario that mirrors real-world construction contract disputes. Imagine a homeowner hires a contractor to remodel their kitchen. The contract specifies high-end materials, but the contractor uses cheaper substitutes to cut costs. The homeowner discovers this after the project is completed and refuses to make the final payment. This is a breach of contract by the contractor, and the dispute arises over the quality of materials used.

The legal issues involved here include breach of contract and misrepresentation. The homeowner could argue that the contractor misrepresented their intentions when they agreed to use high-end materials and then used cheaper substitutes. The contractor, on the other hand, could argue that the materials used were of comparable quality and that the contract did not specify particular brands or suppliers.

The resolution of this dispute could involve negotiation, mediation, arbitration, or litigation. The parties might negotiate a reduction in the final payment to account for the cheaper materials. If they can't agree, they might turn to a neutral third party (a mediator or arbitrator) to help resolve the dispute. If all else fails, the homeowner could sue the contractor for breach of contract.

Contract law plays a crucial role in managing risks in construction projects. A well-drafted contract can anticipate potential issues and provide solutions, reducing the risk of disputes. For example, it can specify the quality of materials to be used, the timeline for the project, and the consequences for delays or substandard work.

Contract modification refers to changes made to the contract after it has been signed. In the construction industry, modifications are common due to the complex and dynamic nature of construction projects. A modification could involve changes to the scope of work, price, or timeline.

Contract modifications are usually handled through change orders, which are written documents that detail the proposed changes and their impact on the contract. Both parties must agree to the change order for it to take effect. It's important to handle contract modifications carefully to ensure that they are legally enforceable and that both parties understand their rights and obligations under the modified contract.

Contract negotiation and execution in the construction industry involve a range of legal and ethical considerations.

From a legal perspective, it's essential to ensure that the contract complies with all relevant laws and regulations. This includes construction codes, labor laws, environmental regulations, and more. The contract should also be clear and unambiguous, with all terms and conditions clearly defined to avoid misunderstandings that could lead to disputes.

Ethically, all parties involved in contract negotiation and execution should act in good faith. This means being honest and transparent about their intentions, capabilities, and constraints. For example, a contractor should not overstate their ability to complete a project within a certain timeframe or understate the cost to win a bid.

Conflicts of interest should also be avoided. For instance, a contractor should not have a financial interest in a supplier or subcontractor without disclosing it to the client.

Fairness is another key ethical consideration. The contract terms should be fair and balanced, not favoring one party excessively over the other. For example, payment terms should be reasonable, and the contractor should be compensated fairly for their work.

Respect for the rights and interests of all parties is also crucial. This includes respecting the client's right to make informed decisions about the project, the contractor's right to fair compensation, and the rights of workers to safe and fair working conditions.

In the construction industry, where projects can be complex and involve significant amounts of money, legal and ethical considerations in contract negotiation and execution are not just good practice—they're essential for the success and reputation of the businesses involved.

Avoiding contract disputes is crucial for smooth project execution and maintaining good relationships with clients. Here are some practical tips for contractors:

1. **Clear Communication:** Ensure all communication with the client is clear and documented. Misunderstandings often lead to disputes, so it's important to ensure everyone is on the same page.
2. **Detailed Contracts:** Make sure your contracts are detailed and cover all aspects of the project, including scope of work, payment terms, timelines, and what constitutes a breach. Vague contracts leave room for interpretation and disputes.
3. **Change Orders:** Any changes to the project should be documented in a change order, which modifies the original contract. Both parties should agree to and sign the change order before the changes are implemented.
4. **Documentation:** Keep thorough records of all aspects of the project, including correspondence, invoices, receipts, and progress reports. This documentation can be invaluable if a dispute arises.
5. **Professional Advice:** Consider seeking legal advice when drafting contracts, especially for large or complex projects. A lawyer can help ensure the contract is legally sound and protects your interests.
6. **Prompt Resolution:** If issues arise, address them promptly. Ignoring problems won't make them go away and could lead to bigger disputes down the line.
7. **Quality Work:** Delivering high-quality work is one of the best ways to avoid disputes. Ensure you have the skills and resources to deliver on your promises.
8. **Honesty and Integrity:** Be honest and transparent with your clients. If problems arise, communicate them immediately and work together to find a solution.

By following these tips, contractors can reduce the likelihood of contract disputes and ensure their projects run smoothly.

Contract law significantly impacts the contractor-client relationship. It provides a legal framework that defines the rights, responsibilities, and expectations of both parties. A well-drafted contract can help build trust and confidence, as each party knows what to expect from the other. Conversely, a poorly drafted contract or a breach of contract can strain the relationship and lead to disputes.

The concept of "good faith" is a fundamental principle in contract execution. It means that both parties should act honestly and fairly, not seek to deceive or take advantage of the other, and cooperate to achieve the contract's objectives. In the construction industry, this could involve the contractor providing accurate estimates, completing the work as promised, and not cutting corners on quality. For the client, good faith could involve providing clear project specifications, making payments on time, and not making unreasonable demands.

Contract law plays a crucial role in dispute resolution in the construction industry. If a dispute arises, the parties will first look to the contract to understand their rights and obligations. The contract might also specify how disputes should be resolved, such as through negotiation, mediation, arbitration, or litigation. If the dispute goes to court, the judge will interpret the contract and apply relevant contract law principles to decide the case.

In essence, contract law provides the rules of the game for the contractor-client relationship. It guides their interactions, shapes their expectations, and provides tools for resolving any disputes that arise. Understanding and applying contract law effectively can help contractors build strong relationships with their clients, deliver successful projects, and protect their rights and interests.

Drafting a construction contract involves several key steps:

1. **Identify the Parties:** Clearly state the names and contact information of the contractor and the client.
2. **Define the Scope of Work:** Detail what work will be done, including the type of work, the location, and any specific methods or materials to be used.
3. **Specify the Price and Payment Terms:** Include the total price, how it was calculated, when payments are due, and what methods of payment are acceptable.
4. **Set the Timeline:** Specify the start and end dates of the project, and any milestones along the way.
5. **Include Terms for Changes and Delays:** Define how changes to the project will be handled, and what happens if the project is delayed due to unforeseen circumstances.
6. **Outline Dispute Resolution Procedures:** Specify how disputes will be resolved, whether through negotiation, mediation, arbitration, or litigation.
7. **Include Termination Clauses:** Define the conditions under which the contract can be terminated by either party.

Legal counsel plays a vital role in contract negotiation and dispute resolution. They can help ensure the contract is legally sound, protect the contractor's interests, and navigate any legal disputes that arise. They can also provide advice on complex legal issues and help negotiate contract terms.

"Force majeure" is a contract law concept that refers to unforeseen events that prevent a party from fulfilling their contractual obligations. These events are beyond the control of the contracting parties and could include natural disasters, war, or government actions. In the construction industry, a force majeure clause in a contract would protect the contractor if they are unable to complete a project due to such events. However, the specific application of force majeure clauses can vary, and it's important to clearly define what constitutes a force majeure event in the contract.

Employment Regulations:

Labor laws are a crucial aspect of the construction industry, affecting how contractors manage their workforce. Here are some key labor laws that contractors should be aware of:

1. **Fair Labor Standards Act (FLSA):** This federal law establishes minimum wage, overtime pay, recordkeeping, and youth employment standards. For example, as of 2023 the federal minimum wage is $7.25 per hour, although many states have higher minimum wages. The FLSA also requires that employees receive overtime pay (1.5 times their regular rate) for hours worked over 40 in a workweek.
2. **Occupational Safety and Health Act (OSHA):** This law requires employers to provide a safe and healthy workplace. Contractors must follow OSHA standards for construction safety, which cover a wide range of issues, from fall protection to hazardous materials.
3. **Workers' Compensation Laws:** These state laws require employers to provide workers' compensation insurance, which provides benefits to workers who suffer work-related injuries or illnesses. The specifics vary by state, but generally, workers' compensation covers medical expenses, a portion of lost wages, and disability benefits.
4. **Family and Medical Leave Act (FMLA):** This federal law allows eligible employees to take unpaid, job-protected leave for specified family and medical reasons. While many construction businesses may be too small to be covered by the FMLA, those with 50 or more employees must comply.
5. **Equal Employment Opportunity Laws:** These laws prohibit discrimination in employment based on race, color, religion, sex, national origin, age, disability, or genetic information. They apply to all aspects of employment, from hiring to firing.

Understanding these labor laws is crucial for contractors. Not only can they help you create a fair and safe workplace, but they can also protect you from legal disputes and penalties. It's also important for the contractor's license exam, as you'll need to demonstrate your understanding of labor laws and their application in the construction industry.

Wage laws play a pivotal role in the construction industry. They ensure that workers are paid a minimum wage and receive overtime pay for working beyond standard hours. As per the Fair Labor Standards Act (FLSA), non-exempt workers must receive at least the federal minimum wage and overtime pay of 1.5 times their regular rate for hours worked beyond 40 in a workweek. However, many states have their own wage laws, often with a higher minimum wage. Non-compliance can result in severe penalties, including fines, back pay for workers, and even criminal charges in extreme cases.

Employment practices in the construction industry cover a broad spectrum. Hiring often involves a thorough review of a candidate's skills, experience, and certifications. Some roles may require specific training or licenses. Firing, on the other hand, must be handled carefully to avoid potential legal issues. Performance management is also crucial. Regular feedback, clear expectations, and objective performance metrics can help ensure that all workers are contributing effectively to the project.

Discrimination laws have significant implications for hiring and management in the construction industry. Under the Civil Rights Act of 1964 and other federal laws, employers cannot discriminate based on race, color, religion, sex, or national origin. Some states have additional protections, such as prohibiting discrimination based on sexual orientation or gender identity. These laws mean that hiring decisions must be based solely on a candidate's ability to perform the job. Similarly, when making management decisions like promotions or layoffs, employers must not consider protected characteristics. Non-compliance can lead to lawsuits, fines, and damage to the company's reputation.

In essence, understanding and complying with wage laws, implementing fair employment practices, and adhering to discrimination laws are not just legal necessities but also contribute to a productive and respectful work environment.

Labor laws significantly influence the contractor-client relationship and project management. They dictate the terms of employment for the contractor's workforce, which can impact project timelines, costs, and quality. For instance, compliance with wage laws and overtime regulations can affect labor costs and scheduling. Safety regulations can influence project procedures and timelines. The contractor's adherence to these laws also affects their reputation with clients and can influence the client's decision to hire or continue working with them.
Labor unions play a significant role in the construction industry. They advocate for workers' rights, negotiate collective bargaining agreements, and can influence wages, hours, and working conditions. Labor laws govern these relationships, outlining the rights and responsibilities of unions and employers. For example, the National Labor Relations Act gives workers the right to form, join, or assist labor organizations and to bargain collectively through representatives of their own choosing.
To ensure compliance with employment regulations, a construction company can follow these steps:

1. **Stay Informed:** Keep up-to-date with federal, state, and local employment laws. This could involve subscribing to relevant newsletters, attending industry seminars, or consulting with legal professionals.
2. **Train Management:** Ensure that all managers and supervisors are trained in employment law compliance. They should understand the laws and how to apply them in day-to-day operations.
3. **Develop Policies:** Create clear, written policies that comply with employment laws. This could include policies on hiring, wage and hour compliance, safety, anti-discrimination, and more.

4. **Implement Procedures:** Develop procedures to implement your policies. This could involve setting up systems to track hours worked and overtime, procedures for handling complaints, or safety protocols.
5. **Document Compliance:** Keep thorough records to demonstrate compliance. This could include time records, safety inspections, training records, and more.
6. **Regularly Review:** Regularly review your policies, procedures, and records to ensure ongoing compliance. This could involve internal audits or external reviews by legal professionals.

By understanding and effectively managing these aspects, a construction company can ensure it complies with employment regulations, thereby reducing the risk of legal issues, improving worker satisfaction, and enhancing its reputation with clients.

Violations of employment regulations can lead to severe legal consequences for construction companies. These can include fines, back pay, compensatory and punitive damages, and even criminal charges in severe cases. Additionally, companies can face reputational damage, which can impact their ability to win future contracts.

Employment regulations also affect subcontractors and independent contractors in the construction industry. While independent contractors aren't covered by many employment laws that apply to employees, misclassifying an employee as an independent contractor can lead to significant penalties. Subcontractors, as separate entities, are responsible for their own compliance with employment laws. However, the main contractor could potentially face indirect consequences for a subcontractor's violations, such as project delays or reputational damage.

The role of state versus federal laws in regulating employment can be complex. Federal laws provide a baseline of protection for workers, but states are free to enact their own laws that provide greater protections. In cases where state and federal laws conflict, generally the law most beneficial to the employee applies.

A real-world example of a legal dispute related to employment regulations in the construction industry is the case of "Walsh Construction Co. v. OSHRC" in 2019. The company was cited by OSHA for a fatal fall at a construction site, alleging that Walsh Construction failed to provide adequate fall protection systems. The company contested the citation, arguing that it had an extensive safety program and that the incident was caused by unpreventable employee misconduct. However, the court upheld the citation, ruling that the company had not done enough to enforce its safety program. This case underscores the importance of not only having safety procedures in place but also ensuring they are effectively implemented and enforced.

Insurance and Bonding:

In the construction industry, various types of insurance are crucial to protect against the inherent risks involved in projects. Here are some of the most relevant types:

General Liability Insurance: This is a fundamental type of insurance that covers bodily injuries and property damage caused by the contractor's operations, products, or injuries that occur on the business premises. For instance, if a passerby is injured due to falling debris from a construction site, general liability insurance would cover the associated costs.
Workers' Compensation Insurance: This type of insurance is mandatory in most states and covers medical expenses and a portion of lost wages for employees who get injured or become ill due to their job. For example, if a worker falls from scaffolding and breaks a leg, workers' compensation insurance would cover their medical bills and some of their lost wages during recovery.
Professional Liability Insurance: Also known as Errors and Omissions (E&O) insurance, this covers legal expenses if a contractor is sued for mistakes (errors) or not doing something they should have done (omissions). For instance, if a design flaw in a building leads to structural damage, professional liability insurance would cover the legal costs and any required reparations.
Liability insurance, in particular, is vital for construction contractors. It serves to protect the financial integrity of the business in the event of unforeseen incidents leading to injuries or property damage. Without it, a single accident could result in substantial out-of-pocket costs, potentially jeopardizing the contractor's business. Moreover, having liability insurance can also enhance a contractor's reputation, as it demonstrates a level of professionalism and preparedness to clients.

Workers' compensation insurance is a type of coverage that provides benefits to employees who suffer job-related injuries or illnesses. In the construction industry, where the risk of physical harm is relatively high, this insurance is particularly crucial. It covers medical care and rehabilitation costs, and it provides disability benefits while the worker is unable to work. It can also provide death benefits to the worker's dependents if the worker dies as a result of a job-related incident. This insurance is not only a legal requirement in many jurisdictions but also a critical component of a responsible employer's risk management strategy.
Turning to bonds, these are guarantees that a specific job will be completed as per the contract terms. In the construction industry, three types of bonds are particularly relevant:

1. **Bid Bonds:** These assure that a contractor will enter into a contract and furnish the required payment and performance bonds if awarded the bid. They protect the project owner if the contractor fails to honor the terms of the bid.
2. **Performance Bonds:** These guarantee that the contractor will perform the work as specified in the contract. If the contractor fails to complete the job or does substandard work, the bond can be used to compensate the project owner or to hire a new contractor to complete the work.

3. **Payment Bonds:** These ensure that the contractor will pay subcontractors, laborers, and material suppliers, protecting those who supply labor or materials for a construction project from non-payment.

Surety bonds serve a critical purpose for construction contractors. They provide a financial guarantee to the project owner, which can be crucial in winning contracts, especially for public or large-scale projects. They also protect the contractor from financial loss if a subcontractor fails to fulfill their contractual obligations. In essence, surety bonds help to foster trust, reliability, and financial security in the often complex and risk-prone construction industry.

Obtaining insurance and bonds in the construction industry typically involves several steps. For insurance, contractors will need to work with an insurance agent or broker who understands the specific risks associated with construction. The process usually involves assessing the contractor's risks, comparing coverage options and rates from different insurers, and then purchasing a policy that meets the contractor's needs and budget.

For bonds, contractors often work with a surety company or a surety bond broker. The surety company will evaluate the contractor's financial strength and track record. This process can include reviewing the contractor's credit history, financial statements, and previous project performance. If the surety company is confident in the contractor's ability to fulfill the contract, it will issue the bond.

Insurance and bonding requirements can vary significantly by state and by project. Different states have different laws and regulations regarding the types and amounts of insurance and bonds that contractors must carry. For example, some states require contractors to have a license bond before they can be licensed, while others do not. Similarly, the size and nature of the project can affect insurance and bonding requirements. Larger, more complex projects typically require higher levels of coverage.

Failing to meet insurance and bonding requirements can have serious consequences for contractors. They may face fines and penalties, lose their license, or be barred from bidding on future projects. If an uninsured or unbonded contractor faces a claim or a lawsuit, they may have to pay out of pocket, which could lead to significant financial hardship or even bankruptcy. Additionally, failing to meet insurance and bonding requirements can damage a contractor's reputation, making it harder to win future contracts.

A real-world example of a contractor's insurance being called upon is the case of a construction company in California that was working on a large commercial building. During the project, a fire broke out due to faulty wiring, causing substantial damage. The construction company's general liability insurance was called upon to cover the costs of the damage, saving the company from potential financial ruin.

Insurance and bonds play a crucial role in the contractor-client relationship and project management. They provide a safety net for both parties. For the client, they offer assurance that the project will be completed as per the contract terms, even if unforeseen issues arise. For the contractor, they provide financial protection against a wide range of risks, from accidents causing injury to workers to damage to the property. This security can foster a more trusting and productive relationship between the contractor and client, and allow for smoother project management.

To ensure adequate insurance and bonding coverage, a construction company can follow these steps:

1. **Assess Your Risks:** Understand the potential risks involved in your projects. This could include worker injuries, property damage, project delays, and more.
2. **Understand the Requirements:** Research the insurance and bonding requirements in your state and for your specific projects. This may involve consulting with a legal expert or an insurance broker.
3. **Compare Providers:** Shop around and compare coverage options and rates from different insurance and surety bond providers. Look for providers who specialize in the construction industry.
4. **Purchase Coverage:** Purchase the insurance policies and bonds that best fit your needs. Make sure to read and understand the terms of each policy or bond.
5. **Regularly Review and Update Coverage:** As your business grows and takes on new projects, your insurance and bonding needs may change. Regularly review your coverage and make updates as necessary.
6. **Maintain Good Records:** Keep detailed records of all your insurance policies and bonds, including policy numbers, coverage amounts, and expiration dates. This will help you stay organized and ensure you're always adequately covered.
7. **Train Your Team:** Make sure your team understands the importance of insurance and bonds and the role they play in your business. This can help ensure everyone is on the same page and reduces the risk of misunderstandings or oversights.

Insurance and bonding play a pivotal role in risk management for construction projects. They act as a safety net, protecting the contractor, the client, and the project itself from a multitude of potential risks.

Insurance policies, such as general liability and workers' compensation, protect against risks like job-related injuries, property damage, and legal liabilities. For instance, if a worker gets injured on the job, workers' compensation insurance can cover their medical expenses and lost wages, protecting the contractor from potential lawsuits. Similarly, if a contractor accidentally causes damage to a client's property, general liability insurance can cover the cost of repairs.

Bonds, on the other hand, protect the financial investment in the project. They ensure that the contractor will fulfill their obligations as per the contract. If a contractor fails to complete a project or fails to pay subcontractors or suppliers, the bond can be used to compensate the project owner or to pay the subcontractors or suppliers.

Subcontractors and independent contractors can also be covered under insurance and bonding policies, but the specifics can vary. In some cases, the general contractor's insurance may cover subcontractors. However, it's often recommended that subcontractors carry their own insurance to ensure adequate coverage. Similarly, subcontractors may be covered under the general contractor's bond, or they may be required to obtain their own bonds, depending on the terms of the contract and the requirements of the project owner.

In essence, insurance and bonding are key tools for managing the inherent uncertainties and risks of construction projects. They provide financial protection and peace of mind for all parties involved, enabling them to focus on successfully completing the project.

Business Management:

In the construction industry, the choice of business structure is pivotal as it influences liability, taxation, and the ability to raise capital. The three most common types are sole proprietorships, partnerships, and corporations.

Sole Proprietorships are the simplest form of business structure. In this setup, one individual owns and operates the business. The advantage of a sole proprietorship is its simplicity and control - you make all the decisions. However, the downside is that the owner is personally liable for all business debts and liabilities.

Partnerships involve two or more people who share the profits and losses of a business. They are relatively easy to establish and offer more financial resources since more than one owner is involved. However, partners are personally liable for business debts, and conflicts between partners can impact the business.

Corporations are more complex structures where the business is a separate legal entity from its owners. This means owners are not personally liable for the company's debts and liabilities. Corporations can also raise capital by selling shares. However, they are more expensive to set up and involve more regulations and tax requirements.

Business planning is crucial in the construction industry. It helps define your business goals, strategies, target market, and financial forecasts. Here's a step-by-step guide to creating a business plan:

1. **Executive Summary:** This is a brief overview of your business, including the name, location, and the product or service you offer.
2. **Company Description:** Detail what your construction company does, the specific services you offer, and the market needs your company fulfills.
3. **Market Analysis:** Research your industry, market, and competitors. Identify trends, themes, and challenges that could impact your business.
4. **Organization and Management:** Describe your business structure, ownership, and the team. Include an organizational chart if necessary.
5. **Services:** Describe in detail the construction services you offer. Explain how these services benefit your customers.

6. **Marketing and Sales Strategy:** Outline your strategies for attracting and retaining customers.
7. **Financial Projections:** Provide an overview of your financial projections, including your expected revenue, expenses, and profitability.
8. **Funding Request (if applicable):** If you're seeking funding, explain how much funding you need and what it will be used for.
9. **Appendix (optional):** An appendix can include resumes of key employees, building permits, contracts, or other supporting information.

Basic accounting principles are vital in the construction industry. They help contractors track income and expenses, making it easier to estimate project costs and ensure profitability. Key principles include:

- **Revenue Recognition:** Revenue should be recognized when earned, not necessarily when received. In construction, this could mean recognizing revenue as different stages of a project are completed.
- **Matching Principle:** Expenses should be recognized in the same period as the revenues they helped generate. For instance, the cost of materials should be recognized as an expense when the project they were used for is billed to the client.
- **Consistency Principle:** Companies should use the same accounting methods from period to period. This makes financial statements comparable over time.
- **Cost Principle:** Transactions should be recorded at their original cost. This means that if you purchase a piece of equipment, it should be recorded at its purchase price.

These principles apply to many financial tasks in construction, such as estimating project costs, creating financial reports, and analyzing profitability.

Let's consider the example of Turner Construction, one of the largest construction management companies in the United States. Turner Construction operates as a subsidiary of the German company Hochtief. This structure allows Turner to leverage the resources of a global parent company while focusing on its core competency of construction management. Turner's business planning strategies include a strong focus on client relationships, risk management, and sustainable building practices.

Financial management is a cornerstone of any successful construction business. It involves planning, organizing, directing, and controlling the financial activities such as procurement and utilization of funds. Understanding basic accounting principles can help contractors manage their finances effectively in several ways:

1. **Budgeting and Forecasting:** By understanding how to read financial statements, contractors can create more accurate budgets and forecasts.
2. **Cash Flow Management:** Understanding the timing of income and expenses can help contractors manage their cash flow, ensuring they have enough money to pay suppliers and employees on time.
3. **Profitability Analysis:** By understanding how to calculate profit margins on individual projects, contractors can identify which types of projects are most profitable and focus their efforts accordingly.

4. **Cost Control:** Understanding how to categorize and track expenses can help contractors identify areas where they can reduce costs.

The choice of business structure can significantly impact a construction company's liability, taxes, and ability to raise capital:

1. **Liability:** Sole proprietors and general partners in a partnership have unlimited personal liability for business debts. In contrast, the owners of a corporation or limited liability company (LLC) are typically not personally liable for business debts.
2. **Taxes:** Sole proprietorships and partnerships typically do not pay income taxes; instead, profits are passed through to the owners, who report the income on their personal tax returns. Corporations, on the other hand, are separate tax entities and pay corporate income taxes.
3. **Ability to Raise Capital:** Corporations and LLCs can sell shares of stock or membership interests to raise capital. In contrast, sole proprietorships and partnerships can only obtain funds through personal contributions, loans, or bringing in additional partners.

In conclusion, understanding these aspects of business management is crucial for anyone preparing for the contractor's license exam. It's not just about knowing how to perform construction work; it's also about understanding how to run a successful construction business.

Strategic planning is a vital component of any successful construction business. It involves setting long-term goals, determining the best approach to achieve those goals, and measuring progress along the way. Here's why it's crucial:

1. **Goal Setting:** Strategic planning helps contractors define clear, measurable goals. These goals might involve expanding into new markets, increasing profitability, or improving customer satisfaction.
2. **Decision Making:** A strategic plan provides a framework for making decisions. When faced with a choice, contractors can refer back to their strategic plan to ensure their decision aligns with their long-term goals.
3. **Performance Measurement:** A strategic plan establishes key performance indicators (KPIs) that contractors can use to measure their progress towards their goals. By regularly reviewing these KPIs, contractors can identify areas where they are performing well and areas where they need to improve.

Now, let's discuss setting up an accounting system for a construction company. Here's a step-by-step guide:

1. **Choose an Accounting Method:** The first step is to choose an accounting method. Most construction companies use either cash or accrual accounting. The cash method is simpler, but the accrual method provides a more accurate picture of a company's financial health.
2. **Set Up a Chart of Accounts:** The chart of accounts is a list of all the accounts in your accounting system. It should include categories for assets, liabilities, equity, income, and expenses.
3. **Implement Job Costing:** Job costing involves tracking all the costs associated with a specific project or job. This includes direct costs like labor and materials, as well as

indirect costs like overhead. Job costing allows contractors to determine the profitability of individual jobs and identify areas where they can reduce costs.

4. **Establish Financial Reporting Procedures:** Regular financial reporting is crucial for monitoring the company's financial health and making informed business decisions. Contractors should establish procedures for generating key financial reports, such as the income statement, balance sheet, and cash flow statement.
5. **Choose Accounting Software:** There are many accounting software options available, ranging from simple spreadsheet programs to comprehensive construction accounting packages. The right choice depends on the size of the company, the complexity of its operations, and its specific accounting needs.
6. **Train Staff:** Finally, it's essential to train staff on how to use the accounting system. This includes not only the office staff who will be entering data and generating reports, but also project managers and other field staff who need to understand the financial aspects of their projects.

Remember, a well-organized accounting system is not just about compliance; it's a tool for managing your business more effectively and profitably.

The construction industry faces several financial challenges, including fluctuating material costs, managing cash flow, and accurately estimating project costs. Understanding basic accounting principles can help contractors navigate these challenges effectively.

Fluctuating Material Costs: Prices for construction materials can vary significantly, affecting project costs and profitability. Contractors need to monitor these costs closely and adjust their pricing strategies accordingly. Understanding the principle of "Cost of Goods Sold" (COGS) can help contractors track how changes in material costs affect their bottom line.

Managing Cash Flow: Construction projects often have long timelines, and contractors may need to pay for labor and materials long before they receive payment from clients. This can create cash flow challenges. Understanding the cash flow statement, a fundamental accounting tool, can help contractors manage their cash inflows and outflows effectively.

Accurate Cost Estimation: Accurately estimating the cost of a project is crucial for setting prices and ensuring profitability. This requires a solid understanding of accounting principles related to cost classification and allocation.

Now, let's discuss some practical tips for managing business finances in the construction industry:

Budgeting: A well-planned budget is a roadmap for your business. It helps you plan for future expenses, make informed business decisions, and measure your performance over time. Start by estimating your income and expenses for a certain period, then track your actual income and expenses and adjust your budget as necessary.

Cash Flow Management: Effective cash flow management ensures that you have enough cash on hand to cover your expenses. This involves closely monitoring your cash inflows and outflows and planning for future cash needs. Strategies might include negotiating longer payment terms with suppliers or scheduling project milestones to align with cash inflows.

Financial Analysis: Regular financial analysis can help you identify trends, spot potential problems, and make informed business decisions. Key financial metrics for construction businesses might include gross margin (sales minus COGS divided by sales), net profit margin (net income divided by sales), and current ratio (current assets divided by current liabilities).

Remember, managing your business finances is not just about keeping the books. It's about using financial information to make strategic decisions that drive your business's success.

Financial Management:

In the construction industry, financial management is a cornerstone of success. A key aspect of this is understanding and utilizing various financial statements. Three primary types of financial statements are particularly relevant:

1. **Income Statements:** Also known as a profit and loss statement, this document provides a summary of a company's revenues, costs, and expenses over a specific period. It helps contractors understand their profitability by showing the net income (revenue minus expenses).
2. **Balance Sheets:** This statement provides a snapshot of a company's financial position at a specific point in time. It lists the company's assets (what it owns), liabilities (what it owes), and equity (the owner's investment). Contractors can use it to assess their financial health and liquidity.
3. **Cash Flow Statements:** This document shows how changes in balance sheet accounts and income affect cash and cash equivalents. It breaks the analysis down to operating, investing, and financing activities. For contractors, it's crucial as it provides insights into their ability to cover payroll and other immediate expenses.

Budgeting, another critical aspect of financial management, is vital for construction projects' success. Here's a basic step-by-step guide to creating a project budget:

1. **Identify Costs:** Start by listing all potential costs. This includes labor, materials, equipment, permits, subcontractor fees, and more.
2. **Estimate Costs:** Use historical data, supplier quotes, and other resources to estimate each cost's value.
3. **Add Contingency:** Construction projects often face unforeseen expenses. Add a contingency line item, typically around 10% of the total estimated cost.

4. **Track and Review:** Once the project begins, track actual costs against the budget. Review and update the budget regularly to reflect changes and maintain financial control.

Tax obligations are another crucial aspect of financial management. Construction contractors may need to pay several types of taxes:

1. **Income Tax:** This is levied on the company's net income. Sole proprietorships and partnerships have their business income taxed as personal income, while corporations pay corporate income tax.
2. **Sales Tax:** Contractors may need to collect and remit sales tax on the services they provide, depending on the state's laws.
3. **Payroll Tax:** If a contractor has employees, they must withhold payroll taxes from employees' wages and pay them to the government. They also need to pay employer payroll taxes.

Understanding these financial management aspects is crucial for contractors to maintain profitability, comply with legal obligations, and ensure the long-term success of their business.

In the construction industry, contractors often face several financial challenges. Here are some of the most common ones and practical tips to overcome them:

1. **Cash Flow Management:** Construction projects often have long payment cycles, which can create cash flow issues. To manage this, contractors can negotiate terms with clients and suppliers to align payment and invoicing schedules. They can also use progress billing, where clients are billed regularly throughout the project, rather than at the end.
2. **Cost Overruns:** Unforeseen issues, changes in scope, or poor cost estimation can lead to costs exceeding the budget. Contractors can mitigate this by improving their cost estimation processes, using contingency reserves, and managing project scope effectively.
3. **Financing:** Contractors often need to invest in materials and labor before receiving payment from clients, which can create financing challenges. To address this, contractors can explore different financing options, such as lines of credit, loans, or supplier credit.

Financial management plays a crucial role in a construction company's profitability and growth. Here are some strategies for improving financial performance:

1. **Cost Control:** Regularly review costs and identify areas for savings. This could involve negotiating better terms with suppliers, improving operational efficiency, or investing in technology to reduce labor costs.
2. **Revenue Enhancement:** Look for opportunities to increase revenue. This could involve expanding into new markets, offering additional services, or improving sales and marketing efforts.
3. **Financial Planning:** Use financial forecasts to plan for the future. This involves estimating future income, costs, and cash flow, and using these estimates to make

informed business decisions. Regularly review and update these forecasts to reflect changes in the business environment.

By understanding and addressing these financial challenges and implementing these strategies, contractors can improve their financial management, leading to increased profitability and growth.

Real-World Example: Bechtel Corporation

Bechtel Corporation, one of the world's most respected engineering, construction, and project management companies, has consistently showcased robust financial management. Their projects span across various sectors, from infrastructure to energy.

Strategies Used by Bechtel:

1. **Diversification:** Bechtel has diversified its portfolio across various sectors and regions. This diversification helps them balance out the risks associated with any one sector or geographic area.
2. **Technological Integration:** The company has integrated advanced technologies into its operations, enhancing efficiency and cost-effectiveness.
3. **Regular Financial Audits:** Bechtel conducts regular financial audits to ensure transparency and to identify areas of improvement.
4. **Risk Management:** They have a dedicated team for risk assessment, ensuring that potential financial risks are identified and mitigated in advance.

Lessons from Bechtel's Experience:

1. **Diversification is Key:** Spreading operations across sectors and regions can help in balancing out risks.
2. **Embrace Technology:** Leveraging technology can lead to cost savings and improved efficiency.
3. **Transparency Matters:** Regular financial audits and transparent operations build trust with stakeholders.

Role of Financial Management in Risk Management:

Financial management plays a pivotal role in risk management for construction projects. By understanding and analyzing financial statements, contractors can gauge their company's financial health, identify potential risks, and make informed decisions.

1. **Budgeting:** A well-prepared budget provides a roadmap for the project. It helps in tracking expenses, ensuring that the project remains financially viable. Any deviations from the budget can be early indicators of potential financial risks.
2. **Cash Flow Analysis:** Regular cash flow analysis ensures that the company has enough liquidity to meet its short-term obligations. A negative cash flow can be a sign of potential financial distress.
3. **Debt Management:** Understanding the company's debt profile can help in assessing its ability to take on more debt or its vulnerability during economic downturns.
4. **Profitability Analysis:** By analyzing profitability ratios, contractors can assess the project's potential return on investment and its overall viability.

In essence, understanding financial statements and budgeting equips contractors with the tools to foresee potential financial pitfalls, allowing them to take proactive measures. Effective financial management not only ensures the project's success but also safeguards the company's reputation and its relationships with stakeholders.

Financial Management Practices by Business Structure:

1. **Sole Proprietorships:**
    - **Simplicity:** Financial management is typically straightforward, as there's no distinction between the business and the owner.
    - **Personal Liability:** The owner is personally liable for all debts. This requires careful financial planning to ensure personal assets are protected.
    - **Taxation:** Income and expenses are reported on the owner's personal tax return, making tax planning crucial.
2. **Partnerships:**
    - **Shared Responsibility:** Financial responsibilities are shared among partners, which can lead to more complex financial management practices.
    - **Joint Liability:** All partners are typically liable for the debts of the business, necessitating clear financial agreements.
    - **Distribution of Profits and Losses:** Financial management must account for the agreed-upon distribution of profits and losses among partners.
3. **Corporations:**
    - **Separate Entity:** Corporations are separate legal entities, leading to distinct financial management practices.
    - **Shareholders:** Financial management must consider dividends and returns for shareholders.
    - **Taxation:** Corporations are taxed separately from their owners, requiring specialized tax planning and management.

Setting Up a Financial Management System for a Construction Company:

1. **Assessment:** Begin by assessing the current financial situation and identifying the company's specific needs.
2. **Choose Accounting Software:** Invest in reliable accounting software tailored for the construction industry, such as QuickBooks for Contractors or Procore.
3. **Set Financial Policies and Procedures:** Establish clear guidelines for expenses, invoicing, payments, and other financial transactions.
4. **Implement Job Costing:** Track expenses and revenues for each project to determine profitability.
5. **Regular Financial Reporting:** Set up monthly or quarterly financial reporting to monitor cash flow, profit and loss, and balance sheets.
6. **Maintain Financial Records:** Keep detailed records of all financial transactions for tax purposes and potential audits.
7. **Review and Adjust:** Periodically review the financial management system and adjust as needed for efficiency and accuracy.

Role of Financial Advisors and Accountants:

1. **Expertise:** They bring specialized knowledge of financial management, tax laws, and industry best practices.
2. **Strategic Planning:** Financial advisors can help contractors plan for growth, investments, and future financial challenges.
3. **Tax Planning and Compliance:** Accountants ensure that contractors comply with tax laws, take advantage of available deductions, and plan for tax liabilities.
4. **Financial Analysis:** They can provide insights into the company's financial health, profitability, and areas for improvement.
5. **Risk Management:** By analyzing financial data, they can identify potential risks and suggest strategies to mitigate them.

In essence, while contractors focus on building and construction, financial advisors and accountants build the financial foundation, ensuring the company's stability and growth.

Trade Skills Section:

This section is crucial because it delves into the technical aspects of construction work, ensuring that contractors are well-equipped to handle the challenges of the job site. Here's an overview:

Trade Skills Overview:

1. **Safety Protocols:**
    - Understanding of OSHA regulations and other safety standards.
    - Proper use of personal protective equipment (PPE).
    - Safe operation of tools and machinery.
2. **Tools and Equipment:**
    - Knowledge of various hand and power tools.
    - Proper maintenance and storage of tools.
    - Selection of the right tool for specific tasks.
3. **Materials Knowledge:**
    - Understanding of different construction materials, their properties, and applications.
    - Proper storage and handling of materials.
    - Best practices for waste management and recycling.
4. **Construction Techniques:**
    - Mastery of foundational construction skills, such as framing, roofing, and masonry.
    - Advanced techniques, like electrical and plumbing installations.
    - Best practices for ensuring quality and durability.
5. **Blueprint Reading:**
    - Ability to interpret and understand architectural and engineering drawings.
    - Knowledge of symbols, scales, and conventions used in blueprints.

6. **Measurement and Estimation:**
    - Accurate measurement techniques for various construction tasks.
    - Skills in estimating material quantities and costs.
7. **Specialized Skills (based on the type of contractor):**
    - For instance, a plumbing contractor would need to know about pipe installations, wastewater systems, and plumbing codes.
    - An electrical contractor would delve into wiring, circuitry, and electrical safety.
8. **Building Codes and Regulations:**
    - Familiarity with local and national building codes relevant to the contractor's specialty.
    - Understanding of permits and inspections.
9. **Sustainability and Green Building:**
    - Knowledge of sustainable construction practices.
    - Familiarity with green building standards and certifications.
10. **Problem Solving and Troubleshooting:**

- Skills to diagnose and rectify common construction issues.
- Techniques for ensuring client satisfaction and addressing concerns.

The "Trade Skills" section is comprehensive, ensuring that contractors are not only knowledgeable but also skilled in the practical aspects of their trade. Mastery of this section is crucial for delivering quality work, ensuring safety, and achieving client satisfaction.

Building Codes and Regulations:

Building codes, at their core, are a set of rules and standards established to ensure the safety, health, and general welfare of building occupants. They touch on everything from structural integrity to fire safety, electrical systems, plumbing, and more.

Historical Evolution of Building Codes: The concept of building codes isn't new. Ancient civilizations, including the Babylonians, had their own versions. The most notable early code is the Code of Hammurabi, dating back to 1754 BC. One of its laws stated that if a builder constructs a house and it collapses, leading to the death of the owner, the builder would be put to death. This underscores the gravity with which construction standards were viewed, even in ancient times.

Fast forward to more recent history, and the Great Fire of London in 1666 was a pivotal event. It led to the establishment of building regulations to prevent future fires. In the U.S., following several devastating fires in the late 19th and early 20th centuries, municipalities began adopting and enforcing building codes to ensure safer construction practices.

Modern building codes, as we recognize them today, began to take shape in the 20th century. In the U.S., for instance, regional codes were developed, such as the Uniform Building Code (UBC) in the West, the Building Officials and Code Administrators (BOCA) in the East, and the Southern Building Code (SBC) in the South. These were eventually unified into the International Building Code (IBC) in the late 1990s, which is now widely adopted and used.

Significance in Construction:

1. **Safety:** The primary purpose of building codes is to protect public health, safety, and general welfare as they relate to the construction and occupancy of buildings and structures.
2. **Standardization:** They provide a standardized method for construction that contractors, architects, and other professionals can follow, ensuring consistency in the quality and safety of buildings.
3. **Economic Stability:** By ensuring buildings are constructed safely, building codes help prevent potential disasters that could result in economic loss.
4. **Innovation and Progress:** As technology and construction methods advance, building codes evolve to accommodate these changes, ensuring that new methods and materials are safe to use.
5. **Legal Protection:** For contractors, following building codes provides a measure of legal protection. If a contractor can demonstrate that they followed the codes in place at the time of construction, it can serve as a defense in legal disputes.
6. **Environmental and Energy Concerns:** Modern building codes also address environmental concerns, promoting energy efficiency and sustainability.

In essence, building codes are the backbone of the construction industry, ensuring that structures are safe, resilient, and built to last. They're a testament to lessons learned from past mistakes and a commitment to protecting future generations.

Understanding the International Building Code (IBC):

The International Building Code (IBC) is a comprehensive, modern building code that addresses the design and installation of building systems through requirements emphasizing performance. It's a model code, meaning it serves as a basis for local jurisdictions to adopt as part of their local codes, often with modifications to reflect local conditions or concerns.

Overview of the IBC:

1. **Origins:** The IBC was introduced in 2000 by the International Code Council (ICC). It was the culmination of efforts to unify three regional codes into a single, unified code that could be adopted and amended by local governments. These three regional codes were the aforementioned Uniform Building Code (UBC), Building Officials and Code Administrators (BOCA), and the Southern Building Code (SBC).
2. **Scope:** The IBC covers all buildings except detached one- and two-family dwellings and townhouses up to three stories, which are covered by the International Residential Code (IRC). It addresses structural safety, fire safety, means of egress, accessibility, energy efficiency, and more.
3. **Updates:** The IBC is updated every three years. This regular revision cycle allows the code to incorporate new technologies, methods, and materials, ensuring it remains current and relevant.

Impact on Construction Standards:

1. **Uniformity:** One of the primary benefits of the IBC is the uniformity it brings. With many jurisdictions adopting it, there's a consistent set of standards across different regions,

making it easier for construction professionals to understand and comply with building requirements.

2. **Safety and Health:** The IBC emphasizes the health and safety of building occupants. It sets standards for fire resistance, structural integrity, ventilation, sanitation, and more, ensuring that buildings are safe environments for their occupants.
3. **Innovation:** The regular update cycle of the IBC means that it's responsive to innovations in the construction industry. New materials and methods can be evaluated and incorporated into the code, ensuring that buildings benefit from the latest advancements while maintaining safety.
4. **Flexibility:** The IBC is performance-based, meaning it often states objectives to be achieved without mandating how to achieve them. This gives architects and builders flexibility in their designs, provided they can demonstrate that their solutions meet the code's requirements.
5. **Economic Impact:** By providing a consistent set of standards, the IBC can lead to cost savings for construction firms that operate in multiple jurisdictions. They don't have to navigate a patchwork of local codes but can work to a consistent set of standards.
6. **Environmental Considerations:** Modern versions of the IBC incorporate standards for energy efficiency and sustainability, reflecting the construction industry's move towards more environmentally friendly practices.

In essence, the IBC plays a pivotal role in shaping the construction landscape, ensuring that buildings are safe, efficient, and up-to-date with the latest industry standards. It's a living document, evolving to meet the needs of a changing world while always prioritizing the safety and well-being of building occupants.

Trade-Specific Codes and Regulations:

1. Electrical:

- **National Electrical Code (NEC):** This is the benchmark for safe electrical design, installation, and inspection. It's used to safeguard people and property from electrical hazards.
- **Key Aspects:** Wiring methods, materials, protection devices, and equipment installation standards.
- **Variations:** Depending on the jurisdiction, local amendments may be added to the NEC, especially in areas prone to natural disasters.

2. Plumbing:

- **International Plumbing Code (IPC):** This code sets standards for the installation and repair of plumbing systems, including water supply, sanitary drainage, and storm drainage.
- **Key Aspects:** Pipe sizing, material usage, fixture requirements, and venting considerations.
- **Variations:** Local codes might have stricter water conservation measures or specific requirements for areas with unique water or soil characteristics.

3. HVAC (Heating, Ventilation, and Air Conditioning):

- **International Mechanical Code (IMC):** This code governs HVAC and refrigeration systems, ensuring they are safely installed and maintained.
- **Key Aspects:** Equipment installation, duct system design, combustion air, chimneys, and ventilation.
- **Variations:** In colder climates, there might be more stringent insulation and heating requirements. In hotter areas, emphasis might be on efficient cooling.

4. Building/Framing:

- **International Building Code (IBC):** This is a comprehensive code that addresses the structural integrity of buildings, including aspects like egress, type of construction, and location on the property.
- **Key Aspects:** Structural design, fire-resistance ratings, and occupancy classifications.
- **Variations:** Seismic zones might have additional requirements for earthquake resistance, while coastal areas might have codes addressing hurricane and flood resistance.

5. Energy:

- **International Energy Conservation Code (IECC):** This code addresses energy efficiency in both residential and commercial buildings.
- **Key Aspects:** Insulation levels, window efficiencies, and HVAC system performance.
- **Variations:** Depending on the climate zone, there might be different insulation and window performance requirements.

6. Accessibility:

- **Americans with Disabilities Act (ADA) Standards:** While not a building code, ADA standards ensure public spaces are accessible to people with disabilities.
- **Key Aspects:** Ramp gradients, door widths, restroom facilities, and parking.
- **Variations:** Some states have their own accessibility codes that might be more stringent than ADA standards.

7. Fire:

- **National Fire Protection Association (NFPA) Codes:** These codes address fire safety in various settings.
- **Key Aspects:** Sprinkler system design, fire alarm systems, and emergency exit pathways.
- **Variations:** Building uses, like whether it's residential, commercial, or industrial, can dictate different fire safety requirements.

Real-World Application: Consider a multi-story commercial building. The electrical contractor must adhere to the NEC for wiring the building, ensuring safe electrical connections. The plumbing contractor references the IPC when installing restrooms on each floor, ensuring proper drainage and water supply. The HVAC contractor, while installing the heating and cooling system, follows the IMC, ensuring efficient and safe climate control. If this building is in a seismic zone, the general contractor must ensure that the building's structural elements meet the IBC's requirements for earthquake resistance. Furthermore, if this building is open to the public, it must also be ADA compliant, ensuring accessibility for all.

In essence, each trade has its own set of codes to ensure safety, functionality, and sometimes even efficiency. While there's overlap, each code addresses the unique challenges and concerns of its respective trade.

Local vs. National Building Codes:

1. Definition:

- **National Building Codes:** These are standardized codes developed by national organizations, aiming to provide a consistent set of standards that can be applied across the country. Examples include the International Building Code (IBC) and the National Electrical Code (NEC).
- **Local Building Codes:** These are codes adopted by local jurisdictions, such as cities or counties. They often start with the national code as a base and then add, modify, or omit certain provisions to better suit the local context.

2. Development and Adoption Process:

- **National Building Codes:** Developed by national committees or organizations, these codes undergo rigorous review processes. They're updated periodically, often every three years, to incorporate new technologies, methods, and safety practices. Once a new version is released, it's up to individual states or local jurisdictions to adopt it.
- **Local Building Codes:** Local jurisdictions decide whether to adopt the national code as-is or modify it. The adoption process usually involves:
    - Review by local building and safety officials.
    - Public hearings or comment periods.
    - Approval by the local governing body, such as a city council or county board.
    - Once adopted, the local code becomes law in that jurisdiction.

3. Differences:

- **Scope:** National codes aim for broad applicability, while local codes can be more specific, addressing unique local conditions like seismic activity, hurricanes, or specific environmental concerns.
- **Flexibility:** National codes provide a baseline. Local codes can be more stringent but usually not less stringent than the national standard.
- **Enforcement:** Local building departments are responsible for enforcing building codes, whether they're based on national standards or have local modifications.

4. Why Differences Exist:

- **Geographical and Environmental Factors:** A coastal city might have stricter codes for wind resistance due to hurricanes, while a city in a seismic zone might have additional requirements for earthquake safety.
- **Historical and Cultural Factors:** Older cities might have provisions related to historic preservation.
- **Economic and Political Factors:** Local economic conditions or political considerations might lead to certain modifications or omissions in the local code.

5. Implications for Contractors:

- **Licensing Exams:** While a contractor's licensing exam might test knowledge of national codes, it's essential to be familiar with local codes when working in a particular jurisdiction.
- **Project Planning and Execution:** Contractors must ensure that their work complies with the local code to avoid costly corrections, delays, and potential legal issues.

Real-World Application: Imagine a contractor based in California, a state known for its seismic activity. While the IBC provides general guidelines on building structures, California has its own set of codes, like the California Building Code (CBC), which incorporates more stringent seismic safety provisions. If this contractor were to work on a project in Florida, they'd need to familiarize themselves with Florida's building codes, which might emphasize hurricane and flood resistance more than seismic safety.

In essence, while national codes provide a consistent baseline, local codes ensure that buildings are suited to their specific environments and contexts. Contractors must always be aware of and comply with the local codes of the jurisdiction in which they're working.

Permits, Inspections, and Code Compliance:

1. Construction Permits:

- **Purpose:** Construction permits ensure that proposed construction work meets local building codes and regulations. They're a formal approval to start construction.
- **Types:** Depending on the jurisdiction and the nature of the work, contractors might need various permits, such as building, electrical, plumbing, mechanical, or demolition permits.
- **Process:** Typically, contractors or property owners submit detailed plans to the local building department. These plans are reviewed for compliance with local codes. Once approved, a permit is issued.

2. Inspections:

- **Purpose:** Inspections verify that the actual construction work complies with the approved plans and local codes.
- **Phases:** Inspections occur at various phases of a construction project:
    - **Foundation Inspection:** Before pouring the foundation.
    - **Rough-In Inspection:** After framing, electrical, plumbing, and HVAC systems are installed but before walls are closed.
    - **Final Inspection:** Upon project completion, ensuring everything is safe and up to code.
- **Outcome:** If the work passes the inspection, the inspector signs off on that phase. If not, they'll provide a list of corrections. Work can't proceed until those corrections are made and re-inspected.

3. Code Compliance:

- **Purpose:** Ensures that structures are safe, energy-efficient, and accessible. It protects the health and safety of the occupants and the public.
- **Enforcement:** Local building departments enforce code compliance through the permit and inspection process.

- **Updates:** Building codes are periodically updated. Contractors must stay informed about the latest changes to ensure compliance in new projects.

4. Consequences of Violations:

- **Stop Work Orders:** If work is being done without a permit or in violation of a permit, the building department can issue a stop work order, halting all construction until the issue is resolved.
- **Fines:** Violations can result in hefty fines, which increase the longer the violation goes unresolved.
- **Liability Issues:** Work that's not up to code can result in safety hazards. If someone gets injured as a result, the contractor could be held liable.
- **Future Sales:** Non-compliant work can cause issues when selling a property. Potential buyers or their lenders might require code violations to be resolved before a sale can proceed.

Real-World Application: Consider a contractor who's renovating a historic home. They start work without obtaining the necessary permits, thinking the changes are minor. However, a neighbor reports the construction to the local building department. An inspector visits the site and finds that not only is the work unpermitted, but it also violates several codes related to electrical wiring and historic preservation. The contractor receives a stop work order and a fine. They must then go through the permit process, make the necessary corrections, and have the work inspected before resuming. This results in delays and additional costs.

In essence, understanding and adhering to the permit, inspection, and code compliance processes are crucial for contractors. Not only do they ensure the safety and quality of construction work, but they also protect contractors from potential legal and financial repercussions.

Energy Codes and Green Building Standards:

1. Energy Codes:

- **Definition:** Energy codes are sets of regulations governing the energy-efficient design and construction of buildings. They set minimum requirements for energy-efficient features such as insulation, windows, lighting, and HVAC systems.
- **Purpose:** The primary goal is to reduce energy consumption, leading to:
    - Lower utility bills for occupants.
    - Reduced strain on the power grid.
    - Decreased greenhouse gas emissions.
- **Examples:** The International Energy Conservation Code (IECC) and ASHRAE Standard 90.1 are two widely recognized energy codes in the U.S.

2. Green Building Standards:

- **Definition:** Beyond energy efficiency, green building standards encompass a broader range of environmental and health considerations, including materials sourcing, indoor air quality, and water efficiency.
- **Purpose:** The objectives are multifaceted:
    - Minimize environmental impact during construction and the building's lifecycle.

    - Create healthier living and working environments.
    - Promote sustainable practices in the construction industry.
- **Examples:** Leadership in Energy and Environmental Design (LEED) and Green Building Initiative's Green Globes are prominent green building certification programs.

3. Significance of Green Building Standards:

- **Environmental Impact:** Green buildings typically have a smaller carbon footprint, use fewer natural resources, and produce less waste.
- **Economic Benefits:** While green buildings might have higher upfront costs, they often result in significant long-term savings due to reduced energy and water consumption.
- **Health and Well-being:** Green buildings prioritize indoor environmental quality, which can lead to better air quality, more natural light, and an overall healthier indoor environment.
- **Market Demand:** As environmental awareness grows, many clients and consumers prioritize green-certified buildings, potentially increasing property values and demand.

4. Integration with Traditional Building Codes:

- **Harmonization:** Many jurisdictions are integrating energy codes and green building standards into traditional building codes, ensuring that all new constructions meet a minimum level of energy efficiency and sustainability.
- **Incentives:** To promote green building practices, some local governments offer incentives such as tax breaks, expedited permitting, or density bonuses for projects that achieve green building certifications.

Real-World Application: Imagine a commercial contractor tasked with constructing an office building in a city that has adopted the IECC and offers incentives for LEED-certified projects. The contractor, aware of the energy code, ensures that the building's insulation, windows, and HVAC system meet the IECC's requirements. Going a step further, they incorporate additional green building strategies, like using recycled materials and installing a green roof. The building not only meets the energy code but also achieves LEED certification. As a result, the building owner benefits from reduced energy costs, qualifies for local tax incentives, and attracts tenants who prioritize sustainability.

In conclusion, as the construction industry evolves and the world grapples with environmental challenges, energy codes and green building standards play a pivotal role in shaping a sustainable and energy-efficient built environment. For contractors, understanding and embracing these standards is not just about compliance but also about staying competitive and meeting the demands of a changing market.

Accessibility, Fire Safety, and Structural Integrity:

1. Accessibility - ADA (Americans with Disabilities Act) Compliance:

- **Definition:** The ADA sets standards to ensure public spaces are accessible to individuals with disabilities.
- **Key Provisions:**
    - **Parking:** Designated handicapped parking spaces with appropriate signage and dimensions.

        - **Entrances:** Ramps, automatic doors, and no-step entrances.
        - **Restrooms:** Accessible stalls, grab bars, and sinks at appropriate heights.
        - **Elevators:** Braille buttons, audio signals, and appropriate door widths.
    - **Implications for Contractors:**
        - Must be familiar with ADA guidelines and ensure all public and commercial spaces they construct are compliant.
        - Non-compliance can result in hefty fines and required modifications.

2. Fire Safety Standards:

- **Definition:** Regulations designed to prevent fire hazards and ensure safe evacuation in the event of a fire.
- **Key Provisions:**
    - **Fire-resistant materials:** Use of non-combustible materials in certain parts of a structure.
    - **Sprinkler systems:** Installation in commercial buildings and multi-family residences.
    - **Fire exits:** Clearly marked, unobstructed paths to safe egress.
    - **Fire alarms:** Properly placed and maintained smoke detectors and alarm systems.
- **Implications for Contractors:**
    - Must be knowledgeable about local fire codes and integrate them into construction plans.
    - Regular inspections often occur to ensure compliance.

3. Structural Integrity:

- **Definition:** Ensuring a building or structure can safely support the loads and forces it will be subjected to during its lifespan.
- **Key Provisions:**
    - **Load-bearing capacities:** Structures must be designed to support dead loads (permanent/static loads like the weight of the building itself) and live loads (temporary/dynamic loads like occupants, furniture, snow).
    - **Seismic considerations:** In earthquake-prone areas, structures must be designed to withstand seismic forces.
    - **Material quality:** Use of materials that meet industry standards for strength and durability.
- **Implications for Contractors:**
    - Must work closely with engineers and architects to ensure structural safety.
    - Regular inspections and tests, like soil tests or load tests, may be required.

Real-World Application: Consider a contractor tasked with building a multi-story office building in California. Due to the state's seismic activity, the contractor must ensure the building can withstand potential earthquakes. This might involve deep foundations, flexible connectors, and specific construction techniques. Additionally, the building must be ADA compliant, with accessible restrooms, ramps, and elevators. Fire safety standards would require the installation of a sprinkler system, clearly marked fire exits, and fire-resistant materials in certain areas.

Failure to adhere to any of these regulations could result in the building not receiving its occupancy permit, legal repercussions, and potential financial losses.
In essence, understanding and adhering to regulations related to accessibility, fire safety, and structural integrity are paramount for contractors. Not only do they ensure the safety and accessibility of the built environment, but they also protect contractors from legal and financial pitfalls.

Updates and Continuing Education:
1. The Dynamic Nature of Building Codes: Building codes aren't static. They evolve in response to new technologies, materials, construction methods, and lessons learned from past construction failures or natural disasters. For instance, after a significant earthquake, building codes might be updated to incorporate newer seismic design techniques.
2. Importance of Staying Updated:

- **Safety:** The primary purpose of building codes is to ensure the safety of the occupants. Being unaware of the latest codes can lead to construction that's not up to current safety standards.
- **Legal Implications:** Non-compliance with the latest codes can result in legal actions, fines, or even the revocation of a contractor's license.
- **Financial Considerations:** A project built not in compliance with the latest codes might need expensive modifications later on.
- **Reputation:** Contractors who are up-to-date are seen as professionals and are more likely to be trusted with significant projects.

3. Continuing Education: Many states and professional organizations require contractors to undergo a certain number of continuing education hours to renew their licenses. These courses ensure contractors are aware of:

- Latest industry best practices.
- New materials and technologies.
- Updated building codes and regulations.

4. Resources for Continuous Learning:

- **Professional Associations:** Organizations like the National Association of Home Builders (NAHB) or the Associated General Contractors of America (AGC) offer courses, seminars, and workshops.
- **Trade Shows and Conventions:** These events often feature seminars on the latest trends, technologies, and codes in the construction industry.
- **Online Platforms:** Websites like the International Code Council (ICC) offer online courses and updates on the latest codes.
- **Local Building Departments:** They often provide resources and training on the latest local code changes.

5. Real-World Application: Imagine a contractor who built homes in the 1990s. Back then, energy efficiency and green building standards might not have been as stringent or even existent in some areas. Fast forward to today, and these aspects are integral to many building

codes. Without continuous learning, this contractor would be ill-equipped to build homes to today's standards, potentially facing legal issues, dissatisfied clients, and financial losses.
In conclusion, the construction industry, like many other sectors, is in a state of continuous evolution. For contractors, staying updated isn't just about compliance; it's about professionalism, ensuring safety, and delivering the best value to clients. Regular updates and continuous education are the tools that ensure contractors remain at the forefront of their trade.

Materials and Methods:

Commonly Used Construction Materials, Their Properties, and Applications:

- **Concrete:** A mixture of cement, water, and aggregates. It's known for its strength and durability. Used in foundations, walls, bridges, and roads.
- **Steel:** An alloy of iron and carbon, it's known for its tensile strength. Used in structural frameworks, reinforcements, and roofing.
- **Brick:** Made from clay, bricks are durable and provide good insulation. They're commonly used in walls, partitions, and facades.
- **Wood:** A natural material that's versatile and provides good insulation. Used in framing, flooring, and decorative elements.
- **Glass:** Used for windows and facades, it allows natural light inside and can provide insulation when used as double glazing.
- **Plastic:** Lightweight and versatile, plastics like PVC are used in piping, insulation, and some cladding.
- **Asphalt:** A sticky, black, and highly viscous liquid or semi-solid form of petroleum. It's primarily used in road construction.

Significance of Material Quality:

Material quality is paramount in construction for several reasons:

- **Longevity:** High-quality materials ensure the structure remains sound for a longer time, reducing the need for repairs or replacements.
- **Safety:** Inferior materials can compromise the structural integrity of a building, posing risks to its occupants. For instance, substandard steel might bend or break under pressure.
- **Cost-Efficiency:** While high-quality materials might be more expensive initially, they can be more cost-effective in the long run due to reduced maintenance costs.
- **Aesthetics:** Quality materials often have a better finish and appearance, enhancing the overall look of the construction.

Environmental Factors Influencing Material Choice:

- **Climate:** In colder regions, materials with good insulation properties like brick or certain woods might be preferred. In humid areas, materials resistant to moisture damage, such as treated wood or certain metals, are essential.
- **Seismic Activity:** In earthquake-prone areas, flexible building materials and techniques that can withstand ground movement are crucial.

- **Soil Type:** The type of soil can determine the foundation type and material. For instance, areas with expansive clay might require specially designed foundations.
- **Availability:** Locally sourced materials can be more environmentally friendly and cost-effective than those transported from far away.
- **Sustainability Concerns:** With growing awareness about environmental issues, there's a push towards using sustainable, recycled, or eco-friendly materials.
- **Regulations:** Some regions have specific environmental regulations that mandate the use of certain materials or prohibit others.

Incorporating these considerations ensures that the constructed building or structure is not only durable and safe but also environmentally responsible and cost-effective.

Essential Tools in the Construction Industry Categorized by Function:

1. **Measuring and Layout Tools:**
   - **Tape Measures:** For taking short to medium-length measurements.
   - **Levels:** To ensure surfaces are horizontal (bubble levels) or vertical (plumb levels).
   - **Theodolites:** For precise angle measurements, typically in large construction or civil engineering projects.
   - **Laser Distance Meters:** For accurate distance measurements over longer spans.
2. **Cutting Tools:**
   - **Hand Saws:** For cutting wood or other soft materials.
   - **Circular Saws:** For cutting wood, plastic, or masonry.
   - **Tile Cutters:** Specifically for cutting tiles.
   - **Bolt Cutters:** For cutting metal bolts or chains.
3. **Fastening Tools:**
   - **Hammers:** For driving nails into surfaces.
   - **Screwdrivers:** For driving screws.
   - **Drills:** Can be used for making holes or driving screws with the right bit.
   - **Nail Guns:** For quickly driving nails into surfaces, especially in framing or roofing.
4. **Digging and Earth Moving Tools:**
   - **Shovels:** For digging and moving materials like soil or gravel.
   - **Pickaxes:** For breaking up hard ground or rock.
   - **Trenchers:** For digging trenches, especially for utilities.
5. **Lifting Tools:**
   - **Cranes:** For lifting heavy materials or equipment.
   - **Hoists:** For lifting or lowering materials using a drum or lift-wheel.
   - **Jack:** For lifting heavy objects a short distance.

Standard Construction Methods from Foundation to Finishing:

1. **Site Preparation:** Before any construction begins, the site is cleared of any vegetation, debris, and potential obstacles. This might involve tree removal, excavation, or leveling.
2. **Foundation Construction:** Depending on the project, various foundation types are used:

    - **Slab-on-Grade:** Concrete is poured directly on a prepared gravel bed.
    - **Crawl Space:** Elevated a few feet above the ground, typically using piers and beams.
    - **Full Basement:** A deeper foundation providing an additional living or storage space below the ground level.
3. **Framing:** This forms the skeleton of the building. Wooden beams and posts are erected to define spaces and support the structure.
4. **Roofing:** Once the frame is up, the roof trusses are installed, followed by sheathing, underlayment, and then the final roofing material, be it shingles, tiles, or metal.
5. **Exterior Walls:** These can be constructed using a variety of materials, including brick, siding, stucco, or stone.
6. **Interior Systems:** This includes the installation of HVAC (heating, ventilation, and air conditioning), electrical wiring, plumbing, and insulation.
7. **Drywall and Interior Surfaces:** Drywall panels are installed, taped, mudded, and sanded. Once primed, they're ready for paint or wallpaper.
8. **Flooring:** Depending on the design, this could be hardwood, tile, carpet, or a combination of materials.
9. **Finishing Touches:** This encompasses a wide range of tasks, from painting walls and installing cabinets to putting in light fixtures and hardware.

Importance of Adhering to Proven Methods: Sticking to established construction methods ensures the safety, durability, and longevity of the structure. Deviating from these methods can lead to structural weaknesses, safety hazards, and potential legal liabilities. Moreover, adherence to proven methods can also streamline the construction process, ensuring projects are completed on time and within budget.

Evolution of Construction Methods: Over time, construction methods have evolved due to:

- **Technological Advancements:** The development of new materials and tools has allowed for faster, more efficient construction.
- **Sustainability Concerns:** There's been a shift towards eco-friendly construction methods and materials.
- **Regulatory Changes:** Building codes and regulations have become stricter, emphasizing safety and energy efficiency.
- **Economic Factors:** Cost-effective methods and materials are continually sought to improve profit margins.
- **Cultural Influences:** Architectural trends and homeowner preferences can influence construction techniques.

The construction industry is a dynamic one, with methods and materials constantly evolving. However, the core principles of safety, efficiency, and durability remain paramount.

Influence of Material Choice on Tools and Methods:

The choice of construction material directly impacts the tools and methods used in a project. Different materials have unique properties, requiring specialized tools for handling, cutting, joining, and finishing.

1. **Handling:** Heavy materials like stone or precast concrete may require cranes, hoists, or other heavy machinery for movement and placement, while lighter materials like timber can often be handled manually or with basic tools.
2. **Cutting and Shaping:** Materials like timber can be cut using saws, while metals might require blowtorches or plasma cutters. Masonry materials, like bricks or stones, might need chisels and masonry saws.
3. **Joining:** Timber can be joined using nails, screws, or adhesives. In contrast, metals might be welded, bolted, or riveted, and masonry materials are typically joined using mortar.
4. **Finishing:** The finishing tools for plaster or drywall differ from those used for timber or masonry. For instance, timber might be sanded and then varnished, while drywall would be taped, mudded, sanded, and then painted.

Real-World Example: The Guggenheim Museum in Bilbao, Spain

Designed by architect Frank Gehry, the Guggenheim Museum is a marvel of modern architecture and construction.

1. **Materials:**
    - **Titanium:** Used for the building's shimmering, undulating facade.
    - **Limestone:** Used for more traditional, orthogonal sections of the building.
    - **Glass:** Large glass curtains provide natural light to the interior.
2. **Tools and Methods:**
    - **Titanium Facade:** Given the complex, free-form design, Gehry and his team used advanced 3D modeling software. The titanium sheets were then cut and shaped using precision tools to fit the unique curves of the structure.
    - **Limestone:** Traditional masonry tools were used for cutting, shaping, and placing the limestone blocks.
    - **Glass:** Cranes and specialized equipment were used to place the large, custom-made glass panels.
3. **Rationale:**
    - **Titanium:** Gehry chose titanium after discovering how its surface gleamed in the rain, making it perfect for Bilbao's often rainy climate. Its malleability allowed for the creation of the building's iconic curves.
    - **Limestone:** It provided a contrast to the futuristic titanium, grounding the building in tradition and echoing other structures in the city.
    - **Glass:** The large glass sections ensure ample natural light, essential for a museum, and provide continuity between the building's interior and the surrounding city.

In essence, the Guggenheim Museum's construction showcases how material choices can dictate the tools and methods used, with each decision playing a crucial role in realizing the architect's vision.

Recent Innovations in Construction Materials and Methods:

1. **Self-Healing Concrete:**

- **Benefits:** This type of concrete contains bacteria that produce limestone when activated by water, effectively "healing" cracks. This can extend the life of structures and reduce maintenance costs.
- **Challenges:** It's more expensive than traditional concrete and may not be suitable for all environments or applications.

2. **3D Printing in Construction:**
    - **Benefits:** 3D printing allows for rapid construction of complex designs, reducing labor costs and waste.
    - **Challenges:** The technology is still in its infancy in terms of large-scale construction. There are concerns about the structural integrity of 3D printed buildings and limitations in the materials that can be used.
3. **Aerogel Insulation:**
    - **Benefits:** Known as "frozen smoke," aerogel is a highly effective insulator, allowing for thinner walls with the same or better insulation properties.
    - **Challenges:** It's more expensive than traditional insulation materials and requires careful handling during installation.
4. **Prefabricated and Modular Construction:**
    - **Benefits:** Building components are manufactured off-site and then assembled on-site, leading to faster construction, reduced waste, and potentially higher quality control.
    - **Challenges:** Transportation of large modules can be challenging. Design flexibility might be limited compared to traditional construction.
5. **Robotics and Automation:**
    - **Benefits:** Robots, like brick-laying or rebar-tying robots, can work continuously, increasing productivity and reducing human-related errors or injuries.
    - **Challenges:** High initial investment and the potential displacement of traditional construction jobs.

Influence on Cost, Timeline, and Overall Project Outcomes:

1. **Cost:**
    - Innovations like self-healing concrete and aerogel insulation might increase initial material costs but can lead to long-term savings due to reduced maintenance or energy costs.
    - Prefabrication, automation, and 3D printing can lead to significant labor cost savings, though they might require higher upfront investments.
2. **Timeline:**
    - 3D printing and prefabricated construction can significantly speed up project timelines by reducing on-site construction time.
    - Robotics and automation can ensure continuous work cycles, further reducing construction timelines.
3. **Overall Project Outcomes:**
    - Advanced materials can enhance the durability and lifespan of structures, leading to better long-term outcomes.

- Innovations like 3D printing allow for more intricate and complex designs, potentially leading to aesthetically superior and functionally optimized buildings.
- Automation reduces human error, potentially leading to higher quality constructions.

In essence, while innovations in construction materials and methods present new challenges, they also offer opportunities for improved efficiency, cost savings, and enhanced project outcomes. As the industry adapts to these changes, the benefits are likely to outweigh the challenges, leading to a more sustainable and efficient construction landscape.

Recent Innovations in Construction Materials and Methods:

1. **Self-Healing Concrete:**
   - **Benefits:** This type of concrete contains bacteria that produce limestone when activated by water, effectively "healing" cracks. This can extend the life of structures and reduce maintenance costs.
   - **Challenges:** It's more expensive than traditional concrete and may not be suitable for all environments or applications.
2. **3D Printing in Construction:**
   - **Benefits:** 3D printing allows for rapid construction of complex designs, reducing labor costs and waste.
   - **Challenges:** The technology is still in its infancy in terms of large-scale construction. There are concerns about the structural integrity of 3D printed buildings and limitations in the materials that can be used.
3. **Aerogel Insulation:**
   - **Benefits:** Known as "frozen smoke," aerogel is a highly effective insulator, allowing for thinner walls with the same or better insulation properties.
   - **Challenges:** It's more expensive than traditional insulation materials and requires careful handling during installation.
4. **Prefabricated and Modular Construction:**
   - **Benefits:** Building components are manufactured off-site and then assembled on-site, leading to faster construction, reduced waste, and potentially higher quality control.
   - **Challenges:** Transportation of large modules can be challenging. Design flexibility might be limited compared to traditional construction.
5. **Robotics and Automation:**
   - **Benefits:** Robots, like brick-laying or rebar-tying robots, can work continuously, increasing productivity and reducing human-related errors or injuries.
   - **Challenges:** High initial investment and the potential displacement of traditional construction jobs.

Influence on Cost, Timeline, and Overall Project Outcomes:

1. **Cost:**

- Innovations like self-healing concrete and aerogel insulation might increase initial material costs but can lead to long-term savings due to reduced maintenance or energy costs.
- Prefabrication, automation, and 3D printing can lead to significant labor cost savings, though they might require higher upfront investments.

2. **Timeline:**
    - 3D printing and prefabricated construction can significantly speed up project timelines by reducing on-site construction time.
    - Robotics and automation can ensure continuous work cycles, further reducing construction timelines.
3. **Overall Project Outcomes:**
    - Advanced materials can enhance the durability and lifespan of structures, leading to better long-term outcomes.
    - Innovations like 3D printing allow for more intricate and complex designs, potentially leading to aesthetically superior and functionally optimized buildings.
    - Automation reduces human error, potentially leading to higher quality constructions.

In essence, while innovations in construction materials and methods present new challenges, they also offer opportunities for improved efficiency, cost savings, and enhanced project outcomes. As the industry adapts to these changes, the benefits are likely to outweigh the challenges, leading to a more sustainable and efficient construction landscape.

Electrical Trade:

Materials:

- Conductors: Copper and aluminum wires used for transmitting electricity.
- Conduits: PVC, metal, or flexible tubing that houses and protects electrical wires.
- Circuit breakers and fuses: Devices that interrupt electrical flow when a fault is detected.

Methods:

- Wiring: Laying out and connecting wires according to electrical plans.
- Circuit design: Planning the flow of electricity to ensure safety and efficiency.
- Grounding: Connecting electrical systems to the earth to prevent electrical shocks.

Plumbing Trade:

Materials:

- Pipes: Copper, PVC, PEX, and galvanized steel are common materials.
- Fixtures: Faucets, sinks, toilets, and showers.
- Valves: Devices that control the flow of water.

Methods:

- Pipe laying: Installing pipes according to plumbing blueprints.
- Leak detection: Using specialized equipment to locate and repair leaks.
- Drainage design: Ensuring water flows away from structures effectively.

HVAC (Heating, Ventilation, and Air Conditioning) Trade:

Materials:

- Ductwork: Galvanized steel or aluminum passages for airflow.
- Refrigerants: Chemicals like R-22 or R-410A used in cooling systems.
- Furnaces and heat pumps: Devices that produce heat.

Methods:

- Load calculations: Determining the heating and cooling needs of a space.
- System design: Planning the layout of HVAC components for optimal efficiency.
- Ventilation: Ensuring proper air exchange to maintain indoor air quality.

Importance of Specialization:

Each trade—electrical, plumbing, and HVAC—requires a deep understanding of its unique materials and methods. Here's why specialization is crucial:

1. **Safety:** Incorrect electrical work can lead to fires. Poor plumbing can result in water damage or health hazards. A faulty HVAC system can lead to carbon monoxide poisoning. Specialized knowledge ensures that work is done safely.
2. **Efficiency:** A specialist knows the best practices for their trade, ensuring that systems run efficiently, saving energy and reducing costs.
3. **Regulations and Codes:** Each trade has specific building codes and regulations. Specialists are trained to know and adhere to these, ensuring that work is up to standard and will pass inspections.
4. **Complexity:** The intricacies of each trade are vast. For instance, an electrician must understand not just wiring, but also load calculations, circuit designs, and grounding techniques. A plumber needs to understand not just pipes, but also water pressure, drainage design, and local water quality issues.
5. **Tools and Equipment:** Each trade uses specialized tools and equipment. Mastery of these tools is essential for high-quality work.

In the construction world, while a general understanding of all trades is beneficial, deep expertise in one area ensures that work is done correctly, efficiently, and safely. This is why many contractors choose to specialize, becoming experts in their chosen field.

The Rise of Sustainable and Eco-Friendly Materials in Construction:

The construction industry has witnessed a significant shift towards sustainability over the past few decades. This change is driven by a combination of environmental concerns, regulatory pressures, and a growing demand from consumers for eco-friendly options.

Benefits of Sustainable and Eco-Friendly Materials:

1. **Environmental Benefits:**
    - **Reduced Carbon Footprint:** Sustainable materials often have a lower carbon footprint in their production, transportation, and installation.
    - **Conservation of Resources:** Many eco-friendly materials are made from renewable resources or have a high recycled content.
    - **Reduced Waste:** Sustainable construction methods often focus on minimizing waste, both in the production of materials and on the construction site.

- **Improved Air Quality:** Many sustainable materials emit fewer volatile organic compounds (VOCs), leading to better indoor air quality.

2. **Economic Benefits:**
    - **Long-term Savings:** While some sustainable materials might have a higher upfront cost, they often last longer and require less maintenance, leading to savings over time.
    - **Energy Efficiency:** Eco-friendly materials, especially in insulation and fenestration, can lead to significant energy savings, reducing heating and cooling costs.
    - **Increased Property Value:** Properties built with sustainable materials and methods can command higher prices in the market.
    - **Tax Incentives and Rebates:** Many governments offer incentives for sustainable construction, which can offset the initial investment.

Methods Incorporating Sustainable Materials:

1. **Green Building Standards:** Systems like LEED (Leadership in Energy and Environmental Design) or BREEAM provide guidelines and certification processes for sustainable construction.
2. **Passive Design:** This architectural method maximizes natural light and heat, reducing the need for artificial lighting and heating. It often incorporates sustainable materials like thermal mass and super-insulation.
3. **Modular Construction:** Building components are manufactured off-site and assembled on-site, reducing waste and increasing efficiency.
4. **Green Roofs and Walls:** These living systems provide insulation, reduce the urban heat island effect, and manage stormwater.
5. **Rainwater Harvesting and Greywater Systems:** These systems reduce the demand on municipal water supplies and manage stormwater on-site.

Long-term Impact on the Construction Industry:

1. **Shift in Demand:** As awareness grows, more clients and developers will demand sustainable options, making it a standard practice rather than a niche.
2. **Regulatory Changes:** Governments worldwide are tightening building codes to address climate change, pushing the industry towards more sustainable practices.
3. **Innovation and New Materials:** The demand for sustainable solutions drives innovation. We're seeing the development of materials like self-healing concrete, low-carbon cements, and bio-based insulation.
4. **Skills and Training:** As the industry evolves, there will be a growing need for skilled professionals trained in sustainable construction methods.
5. **Resilience and Adaptability:** Sustainable construction often considers future climate scenarios, leading to buildings that are more resilient to changing environmental conditions.

In conclusion, the move towards sustainable and eco-friendly materials and methods in construction is not just a trend—it's a necessary evolution. Embracing these changes will lead to a construction industry that's more resilient, efficient, and in harmony with the environment.

Challenges in Sourcing Materials:

1. **Supply Chain Disruptions:** Events like natural disasters, strikes, geopolitical tensions, or pandemics can disrupt the supply chain. For instance, a hurricane might halt the production of a specific wood type in a region, leading to shortages.
2. **Quality Inconsistencies:** Different batches of materials might have variations in quality. For example, bricks from one batch might be slightly different in color or strength than those from another batch.
3. **Price Volatility:** Prices for materials can fluctuate due to demand and supply dynamics, geopolitical events, or changes in regulations.
4. **Regulatory and Environmental Restrictions:** Some materials might be restricted or banned due to environmental or health concerns, leading to sourcing challenges.
5. **Transportation Issues:** Delays in transportation, whether due to logistical issues, customs hold-ups, or other reasons, can lead to materials not arriving on time.

Impact on Construction Methods:

1. **Project Delays:** If a specific material is delayed, it can halt progress, especially if that material is foundational to the subsequent steps.
2. **Cost Overruns:** Delays or having to source alternative (often more expensive) materials can increase project costs.
3. **Compromised Structural Integrity:** Using a lower-quality substitute material might affect the durability and safety of the construction.
4. **Aesthetic Differences:** Variations in material quality can lead to aesthetic inconsistencies in the finished project.

Mitigation Strategies:

1. **Diversified Suppliers:** Don't rely on a single supplier. Having multiple suppliers ensures that if one faces issues, you have alternatives.
2. **Stockpiling Key Materials:** For materials that are crucial and have a known consistent demand, consider maintaining a buffer stock.
3. **Quality Checks:** Implement rigorous quality checks for every batch of material received. This can help in identifying and rectifying inconsistencies early on.
4. **Flexible Construction Methods:** Adopt construction methods that can accommodate slight variations in materials. For instance, if one type of screw is unavailable, ensure that the method allows for an alternative type without compromising on quality.
5. **Stay Updated on Market Trends:** Regularly monitor the market for potential disruptions or price fluctuations. This can help in anticipating challenges and making informed decisions.
6. **Insurance:** Consider insurance options that cover losses due to supply chain disruptions.
7. **Build Strong Relationships with Suppliers:** A good relationship with suppliers can lead to better terms, priority during shortages, and more reliable service.
8. **Continual Learning:** Encourage teams to attend workshops, seminars, and courses (like those for the contractor's license) to stay updated on best practices in material sourcing and risk mitigation.

In the world of construction, the adage "expect the unexpected" holds. By understanding potential challenges and proactively planning for them, contractors can ensure that their projects remain on track, within budget, and uphold the highest quality standards.

Blueprint Reading:

Blueprints are the foundational documents used to translate architectural ideas into tangible structures. They are detailed plans that provide instructions for constructing a building, and they encompass various elements that, when combined, offer a comprehensive view of a construction project. Let's delve into these fundamental elements:

1. **Symbols:**
    - **Purpose:** Symbols are standardized visual representations used to indicate specific components or instructions without the need for lengthy explanations.
    - **Examples:**
        - A zigzag line might represent electrical grounding.
        - A circle with a letter inside could denote a specific type of light fixture.
        - A wavy line might indicate insulation material.
    - **Significance:** Symbols allow for a universal understanding. Regardless of language or regional differences, construction professionals worldwide can interpret these symbols, ensuring consistency and accuracy in construction.
2. **Scales:**
    - **Purpose:** The scale provides a ratio that relates the dimensions on the blueprint to the actual size of the object in the real world.
    - **Examples:**
        - A scale of 1:100 means that for every unit on the blueprint, it represents 100 of the same units in real life.
    - **Significance:** Scales ensure that the entire project can fit on a manageable size of paper while still maintaining accuracy. Contractors use the scale to make precise measurements from the blueprint, ensuring that the actual construction matches the design intent.
3. **Dimensions:**
    - **Purpose:** Dimensions provide specific measurements for various components of the design.
    - **Examples:**
        - The length, width, and height of rooms.
        - The size and placement of windows and doors.
        - The depth and layout of foundations.
    - **Significance:** Dimensions are crucial for accuracy. They ensure that each part of the construction is built to the exact size and location specified by the designer or architect. This ensures structural integrity, functionality, and aesthetic alignment with the design vision.

When combined, these elements offer a **comprehensive view** of a construction project:

- They provide a **visual representation** of what the finished project will look like.
- They offer **detailed instructions** to the construction team, ensuring everyone is aligned in their tasks.
- They ensure that the finished structure is safe, functional, and in accordance with the design intent.

In essence, blueprints serve as a bridge between the conceptual design envisioned by architects and the tangible structure built by contractors. The clarity and detail they provide are paramount to the successful completion of any construction project.

Blueprints are the lifeblood of construction projects, providing intricate details and perspectives that guide the construction process. Let's delve into the different types of blueprints and their unique perspectives:

1. **Floor Plans:**
    - **Description:** A floor plan is a bird's-eye view of a building, showcasing the layout of rooms, hallways, doors, windows, and stairs. It's essentially a map of a level of the structure.
    - **Unique Perspective:** Floor plans give an overhead view of the space layout, allowing contractors to understand the flow of rooms, placement of structural elements, and the relationship between different spaces.
    - **Application:** They're crucial for determining space utilization, traffic flow, and ensuring that the design meets the functional needs of the occupants.
2. **Elevation Views:**
    - **Description:** Elevation views are vertical, flat views of the exterior (or sometimes interior) faces of a building. Think of it as if you're standing in front of a building and looking straight at it. Common elevation views include front, rear, and side elevations.
    - **Unique Perspective:** Elevations provide a clear view of the building's height, the design and size of windows and doors, and other exterior architectural features.
    - **Application:** They're essential for understanding the aesthetic design of the building and ensuring that the exterior aligns with design and zoning requirements.
3. **Section Views:**
    - **Description:** A section view is a cut-through representation of a building, showing its internal structure. Imagine slicing through a building and then looking at it from the side to see its internal components.
    - **Unique Perspective:** Section views reveal the internal construction of the building, including wall thickness, ceiling heights, floor thickness, and sometimes even the materials used.
    - **Application:** They're vital for understanding the structural integrity of the building, ensuring proper insulation, and verifying that internal systems like plumbing and electrical are correctly placed.
4. **Detail Views:**

- **Description:** These are zoomed-in views of specific parts of a construction, providing intricate details that might not be visible in other views.
- **Unique Perspective:** Detail views offer a close-up look at specific construction elements, ensuring that complex components are constructed accurately.
- **Application:** They're often used for areas that require special attention or have intricate designs, such as custom-made fixtures, specialized joints, or unique architectural features.

5. **Site Plans:**
   - **Description:** A site plan is an aerial view of the entire construction site, including the building, landscaping, parking, and other exterior features.
   - **Unique Perspective:** It provides an overview of how the building fits into the entire site, ensuring proper placement, accessibility, and compliance with local zoning regulations.
   - **Application:** Essential for site preparation, landscaping, and ensuring that external utilities and features are correctly placed.

Each type of blueprint offers a unique lens through which contractors can view and understand different facets of the construction project. By combining these perspectives, contractors can ensure that every aspect of the building, from its aesthetic design to its structural integrity, aligns with the project's goals and specifications.

Blueprint symbols and notations act as a universal language in construction, ensuring that everyone involved in a project understands the design intent, regardless of their native language or specific trade expertise. Let's delve into the common symbols and notations and how they vary among electrical, plumbing, and architectural blueprints:

Architectural Blueprints:

1. **Doors and Windows:** Typically represented by straight lines and arcs. The arc indicates the direction in which the door swings open.
2. **Walls:** Represented by thick parallel lines.
3. **Stairs:** Shown as a series of rectangles. The direction of the rectangles indicates whether the stairs ascend or descend.
4. **Breaks:** Zigzag lines indicate that a portion of the structure or component is not shown in the drawing.
5. **North Arrow:** Indicates the orientation of the building in relation to true north.

Electrical Blueprints:

1. **Outlets:** Represented by a small circle or rectangle, sometimes with lettering to indicate the type (e.g., "S" for a switch or "R" for a receptacle).
2. **Light Fixtures:** Depicted by various symbols, such as a circle with a line for a ceiling light or a wall sconce symbol for wall-mounted lights.
3. **Switches:** Typically shown as a break in a line with an arc over the top.
4. **Circuit Breakers and Panels:** Represented by a series of lines or rectangles, often numbered to indicate circuits.

5. **Wiring:** Lines connecting different electrical components. Dotted or dashed lines might indicate wiring that's hidden behind walls.

Plumbing Blueprints:

1. **Pipes:** Represented by straight lines, with different line types (e.g., solid, dashed) indicating the type of pipe or its location (above or below the floor).
2. **Valves:** Shown as small rectangles or circles, often with lettering to indicate the type (e.g., "GV" for gate valve).
3. **Fixtures:** Symbols for sinks, toilets, bathtubs, and other fixtures are standardized but can vary slightly based on the specific design.
4. **Hot and Cold Water:** Often indicated by red and blue lines or the letters "H" and "C".
5. **Drain, Waste, and Vent (DWV):** These systems might have unique symbols, especially for components like traps or vents.

It's essential to note that while these symbols are relatively standardized, variations can occur based on the region, the drafting professional, or the specific project. Therefore, every set of blueprints should come with a legend or key that deciphers all the symbols used in that particular set of drawings. For those preparing for the contractor's license exam, familiarizing oneself with these symbols and their variations across trades is crucial, as they form the foundation of understanding construction plans and executing projects accurately.

In the realm of construction, the concept of scale in blueprints is of paramount importance. It ensures that the vast expanse of a project can be captured on a manageable sheet of paper, while still retaining every minute detail.

Scale in Blueprints: The scale is a ratio that defines the relationship between the dimensions on the blueprint and the actual dimensions of the structure. For instance, a scale of 1:50 means that for every one unit on the blueprint, it represents 50 of the same units in real life. So, if a wall is drawn as 2 inches long on the blueprint, it would be 100 inches (or 8.33 feet) in reality.

Ensuring Accurate Measurements:

1. **Using Architect's Scales:** Contractors use specialized rulers known as architect's scales. These are not your typical rulers; they have multiple units of measure to correspond with common blueprint scales. By selecting the correct scale, contractors can directly measure the blueprint and get accurate real-world dimensions.
2. **Consistent Annotations:** Blueprints often come with specific measurements annotated. Contractors must always cross-reference these with scaled measurements to ensure consistency.
3. **Digital Tools:** Modern technology has blessed contractors with digital tools and software that can automatically scale and measure blueprints. These tools can instantly provide real-world dimensions, reducing the risk of human error.
4. **Triple-Check:** It's a golden rule in construction – measure thrice, cut once. Before making any irreversible changes, contractors often verify measurements multiple times.
5. **Calibration:** If using digital devices like laser measures, it's crucial to calibrate them regularly to ensure their accuracy.

Translating Blueprint into Physical Structure:

1. **Layouts and Markings:** Before any construction begins, contractors mark out the site. This involves translating every wall, window, door, and other structural elements from the blueprint to the ground, ensuring everything is in the right place and of the right size.
2. **Regular Verification:** As construction progresses, contractors will consistently refer back to the blueprint, verifying that the work aligns with the plan.
3. **Collaboration:** It's not just the responsibility of one individual. Site supervisors, architects, and other specialists will often collaborate, cross-referencing their work with the blueprint to ensure accuracy.
4. **Adjustments:** Sometimes, real-world challenges mean slight deviations from the blueprint are necessary. In such cases, it's crucial to understand the implications of any change on the overall structure and to make adjustments that are in line with safety and design integrity.

In the context of the contractor's license exam, understanding the concept of scale and its practical application is vital. It's not just about knowing what scale is, but understanding its significance in ensuring that a structure is built accurately, safely, and to the client's specifications.

Blueprints, while pivotal, are just one piece of the comprehensive documentation puzzle in construction. Let's delve into the other essential construction documents that accompany blueprints:

1. Specifications (Specs):

- **Purpose:** These are detailed written documents that define the quality and type of materials, products, and workmanship required for the project.
- **Context:** While a blueprint might show where a door is located, the specifications will detail the type of wood, the finish, the type of handle, and even the installation method.

2. Schedules:

- **Purpose:** Schedules provide a timeline for the project, detailing when each phase or task should start and finish.
- **Context:** If a blueprint shows the layout of a three-story building, the schedule will indicate when foundation work should complete, when the first floor should be erected, and so on.

3. Contracts:

- **Purpose:** This is a binding agreement between parties, typically the owner and the contractor. It outlines the scope of work, payment details, responsibilities, and other legalities.
- **Context:** While blueprints and specifications detail the "what" and "how" of the project, the contract outlines the "who," "when," and "how much."

4. Addenda:

- **Purpose:** These are changes or clarifications made to the original set of bidding documents before the bidding is closed. They become part of the contract documents.

- **Context:** If there's a change in the blueprint after it has been issued to bidders, an addendum will detail that change.

5. Change Orders:

- **Purpose:** After the contract is executed, any changes to the work or price are documented as change orders. They detail any added or deducted work and the associated cost.
- **Context:** If, during construction, a wall's location shown on the blueprint needs to be moved, a change order would document this shift and any associated cost changes.

6. Shop Drawings:

- **Purpose:** These are detailed drawings created by contractors, suppliers, or manufacturers to show the specific fabrication or installation details.
- **Context:** A blueprint might indicate a custom window's location, but the shop drawing will show its exact dimensions, materials, and installation details.

7. As-Built Drawings:

- **Purpose:** These are revised blueprints updated to reflect all changes made during the construction process.
- **Context:** Once construction is complete, as-built drawings provide a record of the exact dimensions, materials, and locations of all elements, as they were actually built.

In essence, while blueprints provide a visual representation of the project, these accompanying documents offer the intricate details, timelines, responsibilities, and changes. Together, they ensure that every stakeholder has a clear understanding of every facet of the project. For the contractor's license exam, having a comprehensive grasp of these documents, their interplay, and their significance in the construction process is crucial.

Let's imagine a real-world scenario:
Project: A single-story residential home.
Blueprint Overview: The blueprint showcases the layout of the house, including the living room, kitchen, two bedrooms, two bathrooms, and a garage.
Interpreting the Blueprint:

1. **Title Block:** In the bottom right corner, there's a box that provides essential details:
    - Project Name: "Smith Residence"
    - Address: "123 Elm Street"
    - Date: "January 1, 2023"
    - Architect: "John Doe & Associates"
    - Scale: "1/4 inch = 1 foot"
2. **Symbols:**
    - **Doors:** Door symbols look like a quarter-circle. The side the arc is on indicates the door's swing direction.
    - **Windows:** Parallel lines with a break in the middle.
    - **Electrical Outlets:** Small circles or dots, sometimes with a letter inside to indicate the type (e.g., "S" for a switch or "R" for a receptacle).

- **Plumbing Fixtures:** Symbols in the bathrooms and kitchen represent sinks, toilets, and bathtubs.

3. **Measurements:**
    - Room dimensions are indicated in feet and inches. For instance, the master bedroom might be labeled "14'x16'," meaning it's 14 feet wide and 16 feet long.
    - Wall thickness is also indicated, often being 4.5" (which includes a standard 2x4 wall and half-inch drywall on both sides).
4. **Annotations:**
    - Notes might be present to provide additional information. For instance, next to the living room's large window, a note might read: "Bay window - refer to window schedule for details."
5. **Elevation Views:**
    - On a separate sheet, there are drawings of the home's exterior from the front, back, and sides. This provides a view of the house's appearance from the outside, indicating window heights, roof pitches, and exterior finishes.
6. **Section Views:**
    - Another sheet might cut through the house, typically from the ground to the roof, showing details like foundation depth, ceiling heights, and roof construction.

Accompanying Documents:

1. **Specifications (Specs):** A detailed document accompanies the blueprint, specifying the type of wood for the flooring, the brand and model of kitchen appliances, the kind of tiles for the bathroom, etc.
2. **Schedules:**
    - **Window Schedule:** Lists all windows by type, size, and location.
    - **Door Schedule:** Details door types, materials, and locations.
3. **Contract:** This would outline the project's cost, timeline, responsibilities of both the homeowner and the contractor, and other legalities.
4. **Addenda or Change Orders:** If there were changes after the initial blueprint and specs were given, these would be documented separately.

In this example, interpreting the blueprint requires understanding the symbols, measurements, and notes provided. The accompanying documents offer further detail, ensuring the contractor knows precisely what materials to use and how the finished project should look. For the contractor's license exam, understanding how to read and interpret these details is crucial.

Estimating and Bidding:

Estimating project costs is a meticulous process that requires a deep understanding of various components to ensure accuracy. Let's delve into these components:

1. Labor:
Labor costs encompass wages for skilled and unskilled workers, benefits, and any additional allowances. It's crucial to account for the number of workers, their respective hourly rates, and the estimated time they'll spend on the project. For instance, if specialized trades like electricians or plumbers are required, their rates might be higher than general laborers.

2. Materials:
This involves calculating the quantity and cost of every material to be used. From concrete to nails, every item's cost is listed. Prices can be sourced from suppliers' quotes or catalogs. It's essential to consider the quality of materials too; higher-grade materials might cost more but can offer better longevity and finish.

3. Equipment:
Whether it's heavy machinery like cranes or specific tools like power drills, equipment rental or purchase costs must be factored in. The duration of equipment usage, maintenance, and fuel (if applicable) are also considered.

4. Overhead:
Overhead costs are indirect expenses that can't be attributed to a specific task but are essential for project completion. This includes utilities, site security, temporary facilities, insurance, and permits. It's a common practice to calculate overhead as a percentage of the total direct costs (labor, materials, and equipment).

5. Profit Margin:
After all costs are accounted for, contractors add a profit margin. This margin compensates for the risk taken and ensures the financial viability of the business. The percentage can vary based on industry standards, project complexity, and the contractor's business strategy.

Determining the Overall Project Estimate:
Once each component is quantified, they're summed up to provide the total project estimate. This estimate gives the contractor and the client a clear picture of the project's financial scope. For example, if a contractor estimates the labor at $50,000, materials at $30,000, equipment at $10,000, overhead at 15% of direct costs ($13,500), and aims for a profit margin of 10% on the total costs ($103,500), the overall project estimate would be $113,850.

Practical Application:

Imagine a small residential project where high-quality wooden flooring is chosen. While the material cost might be high, the labor cost might decrease due to the ease of installation compared to a more complex material. However, the equipment needed for precise cutting might increase the equipment cost. By understanding these interdependencies, contractors can make informed decisions, ensuring the project remains within budget while meeting quality standards.

In conclusion, accurate cost estimation is a blend of science and art. While the science involves precise calculations and data, the art lies in forecasting uncertainties, understanding market trends, and leveraging experience from past projects. For those preparing for the contractor's license exam, mastering this balance is pivotal.

Preparing and submitting a bid is a comprehensive process that requires a blend of technical knowledge, market understanding, and strategic thinking. Here's a step-by-step breakdown:

1. Bid Solicitation: Before anything else, contractors need to be aware of projects up for bidding. This can be through public notices, industry publications, or direct invitations from clients.

2. Review Project Documents: Once a project is identified, contractors review the provided documents, which typically include blueprints, specifications, and other related materials. This helps in understanding the project's scope and requirements.

3. Site Visit: If possible, visiting the construction site can provide valuable insights. It helps in identifying potential challenges like accessibility, ground conditions, and surrounding infrastructure.

4. Cost Estimation: This is a crucial step where contractors estimate costs for labor, materials, equipment, overhead, and other project-specific expenses. It's essential to be thorough and accurate to avoid underestimating or overestimating costs.

5. Determine Profit Margin: On top of the estimated costs, contractors decide on a profit margin. This is based on factors like project complexity, competition, and potential risks.

6. Prepare the Bid Proposal: The bid proposal is a formal document that outlines the contractor's offer. It includes the total cost, project timeline, methods, and other relevant details. Some bids might also require a proposed project schedule and methodology.

7. Review and Submit: Before submission, it's crucial to review the bid for accuracy and completeness. Once confident, the bid is submitted before the deadline.

8. Post-Bid Meeting: Some clients might hold meetings after bid submission to clarify or negotiate certain aspects of the proposal.

Ensuring a Competitive Yet Profitable Bid:

- **Market Research:** Understanding the current market rates and what competitors might offer is key.
- **Efficiency:** Leveraging technology, past experience, and industry best practices can lead to cost savings without compromising quality.
- **Supplier Relationships:** Building good relationships with suppliers can lead to discounts or better terms.

Common Pitfalls to Avoid:

- **Overlooking Details:** Missing out on any project detail can lead to underestimation of costs.
- **Overcommitting:** Promising unrealistic timelines or prices can harm the contractor's reputation and profitability.
- **Not Reviewing:** Failing to double-check the bid can lead to errors or omissions.
- **Ignoring External Factors:** Not considering potential external factors like weather conditions, market fluctuations, or regulatory changes can impact project costs and timelines.

In essence, successful bidding is a balance between being competitive to win the project and ensuring profitability. It's a skill that contractors refine over time, learning from each bid, whether successful or not. For those preparing for the contractor's license exam, understanding this process and its nuances is fundamental.

The construction industry has seen a significant shift in the tools and methodologies used for estimating, largely due to technological advancements. Here's a deep dive into the modern tools and their impact:

1. Construction Estimating Software: Modern estimating software allows contractors to input project details and automatically calculate costs based on pre-set parameters. Examples include ProEst, PlanSwift, and Bluebeam. These tools:

- **Streamline the Estimation Process:** By automating calculations, reducing manual data entry.
- **Improve Accuracy:** By reducing human error and using up-to-date pricing databases.
- **Facilitate Collaboration:** Multiple team members can work on an estimate simultaneously.

2. Building Information Modeling (BIM): BIM is a 3D model-based process that gives construction professionals insights and tools to efficiently plan, design, construct, and manage buildings. Tools like Autodesk's Revit and Navisworks allow for:

- **Detailed Visualization:** BIM provides a detailed 3D model of the project, helping estimators visualize and understand project scope better.
- **Integration with Estimating:** Directly linking the model to estimating tools, ensuring that any design changes automatically update the estimate.

3. Digital Takeoff Tools: Gone are the days of manual takeoffs with paper blueprints. Digital tools allow estimators to measure lengths, areas, and volumes directly from digital blueprints. This:

- **Speeds Up the Process:** Digital measurements are faster than manual ones.
- **Enhances Precision:** Reducing the chances of human error.

4. Cloud-Based Collaboration: Platforms like Procore or Buildertrend facilitate real-time collaboration. Estimators, project managers, and site supervisors can all access the same data, ensuring consistency.

5. Databases and Price Libraries: Modern estimating tools often come with integrated price libraries, ensuring that estimates are based on the latest material and labor costs. These databases can be region-specific, ensuring localized accuracy.
6. Mobile Applications: With the rise of smartphones and tablets, many estimating tools have mobile versions. This allows for on-the-go estimating, especially useful during site visits.
Technological Advancements' Impact on Estimating:

- **Increased Accuracy:** Automated tools and up-to-date databases reduce the chances of errors.
- **Efficiency:** What once took days can now be done in hours.
- **Consistency:** Standardized tools and methods ensure that estimates are consistent across projects.
- **Data-Driven Decisions:** With access to historical data and analytics, contractors can make more informed decisions.

Real-World Example: Consider a contractor using BIM for a commercial building project. The 3D model allows them to visualize the entire structure, including the electrical, plumbing, and HVAC systems. By integrating this model with an estimating tool, they can automatically calculate the quantity of materials required, like the number of bricks or the length of electrical wire. If the design changes, the estimate updates automatically. This level of integration ensures that the estimate is always aligned with the design, reducing the chances of costly overruns.
For those preparing for the contractor's license exam, understanding these modern tools and methodologies is crucial. Not only do they represent the future of the industry, but they also play a significant role in ensuring project success.

Estimating is a critical phase in the construction process, and while modern tools and methodologies have improved accuracy, there are inherent risks and uncertainties. Here's a deep dive into these challenges and strategies to address them:
Risks and Uncertainties in Project Estimating:

1. **Fluctuating Material Prices:** Prices for materials can vary due to market demand, geopolitical events, or supply chain disruptions.
2. **Labor Costs:** Estimating labor can be tricky, especially if there's a shortage of skilled workers or if union negotiations change wage rates.
3. **Site Conditions:** Unforeseen site conditions, like contaminated soil or hidden rock formations, can lead to increased costs.
4. **Design Changes:** Changes in project design or scope can significantly impact the estimate.
5. **Regulatory and Permitting Issues:** Delays or changes in obtaining necessary permits can lead to increased costs.
6. **Economic Factors:** Economic downturns or inflation can impact project costs.
7. **Equipment Availability and Costs:** The availability of specific machinery or its operational costs can vary.

Strategies to Account for and Mitigate Risks:

1. **Contingency Funds:** Always include a contingency amount in the estimate. This is a percentage of the total estimated cost set aside to cover unforeseen expenses. The exact percentage can vary, but 10-15% is common for many projects.
2. **Detailed Site Analysis:** Before estimating, conduct a thorough site analysis. This can help identify potential issues like soil conditions or accessibility challenges.
3. **Regularly Update Price Databases:** Ensure that the pricing databases or libraries used in estimating tools are updated regularly to reflect current market conditions.
4. **Clear Communication with Design Team:** Regular communication can help identify potential design changes early, allowing for adjustments in the estimate.
5. **Phased Estimating:** As the project progresses and more details become available, refine and update the estimate. This iterative process can increase accuracy.
6. **Risk Analysis:** Identify potential risks and assign a probability and impact score to each. This can help prioritize which risks to address first.
7. **Alternative Materials and Methods:** Consider alternative materials or construction methods that might offer cost savings or reduce risk.
8. **Fixed-Price Contracts:** For some materials or services, consider locking in prices through fixed-price contracts to guard against price fluctuations.
9. **Stay Informed on Regulatory Changes:** Regularly review local, state, and federal regulations to anticipate potential changes that could impact costs.
10. **Engage Experts:** For specialized parts of the project, consider consulting with experts. For instance, if there's a complex electrical component, an electrical engineer's input can provide a more accurate estimate.

During the Bidding Phase:

1. **Thorough Bid Review:** Before submitting, review the bid multiple times to ensure all aspects of the project have been considered.
2. **Seek Clarifications:** If there are ambiguities in the project documents, seek clarifications rather than making assumptions.
3. **Competitive Subcontractor Bids:** If using subcontractors, get multiple bids to ensure competitive pricing.
4. **Transparency with Clients:** Be upfront about potential risks and how they've been accounted for in the bid. This can build trust and set clear expectations.

In conclusion, while estimating will always involve some level of uncertainty, a systematic approach, combined with modern tools and methodologies, can significantly mitigate risks. For those preparing for the contractor's license exam, understanding these strategies is crucial, as they form the foundation of successful project management in the construction industry.

Project Management:

Core Principles and Methodologies of Managing a Construction Project:

1. **Initiation:**
    - **Needs Identification:** Before any construction project begins, there's a need or a problem that requires a solution. This could be the need for additional office space, a new bridge, or a residential complex.
    - **Feasibility Studies:** Once the need is identified, studies are conducted to determine the project's viability. This includes assessing financial, technical, and environmental aspects.
2. **Planning:**
    - **Scope Definition:** Clearly define what the project will achieve, outlining specific goals, deliverables, features, functions, tasks, deadlines, and costs.
    - **Resource Allocation:** Determine the resources (labor, equipment, materials) required and allocate them appropriately.
    - **Risk Management:** Identify potential risks and develop strategies to mitigate them.
    - **Budgeting:** Estimate the costs for the entire project and allocate funds accordingly.
    - **Scheduling:** Develop a timeline for the project, breaking down tasks and allocating time for each.
3. **Execution:**
    - **Team Mobilization:** Assemble the team, provide necessary training, and allocate tasks.
    - **Resource Utilization:** Ensure resources are used efficiently and effectively.
    - **Stakeholder Communication:** Maintain open lines of communication with all stakeholders, ensuring they are informed and involved as necessary.
4. **Monitoring:**
    - **Performance Tracking:** Regularly track and measure project performance against the plan. Use tools like Earned Value Management (EVM) to gauge project health.
    - **Quality Control:** Implement regular checks and inspections to ensure work quality meets or exceeds standards.
    - **Adjustments:** If deviations from the plan occur, adjustments are made to bring the project back on track.
5. **Closure:**
    - **Project Evaluation:** Once the project is complete, evaluate its success against the original objectives.
    - **Documentation:** Ensure all project documentation is completed, including contracts, financials, and other relevant information.
    - **Feedback and Lessons Learned:** Gather feedback from stakeholders and document lessons learned for future projects.

How These Principles Ensure Success:

- **Holistic View:** By breaking down the project into distinct phases, managers can focus on each part's specific requirements, ensuring nothing is overlooked.
- **Risk Mitigation:** Through planning and monitoring, potential issues are identified early, allowing for proactive solutions rather than reactive fixes.
- **Resource Optimization:** Proper planning ensures that resources, whether human, material, or financial, are used optimally, preventing wastage and ensuring efficiency.
- **Stakeholder Satisfaction:** Regular communication with stakeholders ensures their needs and concerns are addressed, leading to better relationships and project outcomes.
- **Continuous Improvement:** The closure phase's feedback and lessons learned provide invaluable insights for future projects, fostering continuous improvement in construction project management.

In essence, these core principles and methodologies provide a structured framework that guides the project from conception to completion, ensuring efficiency, effectiveness, and stakeholder satisfaction.

Gantt Charts:

- **What it is:** A horizontal bar chart that represents a project schedule over time.
- **How it works:** Tasks or activities are listed vertically, and the timeline spans horizontally. Each task has a corresponding bar that represents its start and end dates.
- **Benefits:**
    - Provides a visual representation of the project timeline.
    - Easily identifies overlapping tasks and task dependencies.
    - Useful for tracking progress and ensuring tasks start and end as planned.
- **Real-world application:** For a home construction project, a Gantt chart might show that foundation work will occur in the first week, followed by framing in the second and third weeks, overlapping with electrical work.

Critical Path Method (CPM):

- **What it is:** A step-by-step technique used to determine the sequence of tasks that form the longest duration, allowing the shortest time to complete the project.
- **How it works:**
    - List all tasks required to complete the project.
    - Note dependencies between tasks.
    - Calculate the earliest and latest each task can start and finish without delaying the project.
- **Benefits:**
    - Identifies the most critical tasks that, if delayed, would delay the entire project.
    - Helps in resource allocation by highlighting crucial tasks.
- **Real-world application:** In building a skyscraper, if the elevator installation (a critical task) is delayed, it could delay subsequent tasks like interior finishing.

Program Evaluation Review Technique (PERT):

- **What it is:** A statistical tool that considers uncertainty and variability in task durations.

- **How it works:**
  - Tasks, durations, and dependencies are identified.
  - Three time estimates are made for each task: optimistic, most likely, and pessimistic.
  - These estimates are used to calculate an expected duration and variance for each task.
- **Benefits:**
  - Provides a more realistic understanding of project completion times and potential delays.
  - Helps in risk assessment by identifying tasks with high variability.
- **Real-world application:** In a dam construction project, unpredictable factors like weather conditions can affect tasks. PERT helps in estimating realistic timelines considering such uncertainties.

How These Tools Aid in Efficient Project Execution:

1. **Visibility:** All these tools provide a clear visual representation of the project's timeline, making it easier for stakeholders to understand the project flow.
2. **Resource Allocation:** By identifying critical tasks (especially with CPM), resources can be efficiently allocated to ensure no delays.
3. **Risk Management:** PERT, with its focus on variability, allows project managers to anticipate and plan for uncertainties.
4. **Performance Tracking:** Gantt charts, in particular, are excellent for tracking project progress against planned timelines.

In the realm of the contractor's license exam, understanding these tools is paramount. Not only do they form the backbone of project management in construction, but they also represent the industry's commitment to efficiency, risk management, and timely delivery. Knowing how to utilize and interpret these tools can significantly impact a contractor's success on the job and on the exam.

Let's delve into the significance of quality control in construction.

Importance of Quality Control in Construction:

1. **Safety:** Ensuring quality means adhering to safety standards, reducing the risk of accidents during construction and ensuring the long-term safety of the structure's occupants.
2. **Durability:** Quality control ensures that structures last for their intended lifespan, reducing the need for costly repairs or premature replacements.
3. **Cost Efficiency:** While quality materials and processes might have a higher upfront cost, they often result in long-term savings by reducing maintenance and repair costs.
4. **Client Satisfaction:** Delivering a project that meets or exceeds client expectations can lead to repeat business and referrals.
5. **Regulatory Compliance:** Quality control ensures adherence to building codes and regulations, avoiding potential legal issues or penalties.

Processes and Best Practices for Quality Control:

1. **Material Selection:**
    - **Testing:** Before use, materials should be tested for properties like strength, durability, and safety.
    - **Certifications:** Only use materials that come with certifications from recognized industry bodies.
    - **Storage:** Ensure materials are stored correctly to prevent degradation.
2. **Workmanship:**
    - **Training:** Ensure that all workers are adequately trained for their specific tasks.
    - **Supervision:** Regularly supervise tasks, especially those that are critical to the project's integrity.
    - **Checkpoints:** Establish regular checkpoints for tasks to ensure they're completed to the required standard before moving on.
3. **Inspections and Audits:**
    - **Regular Inspections:** Schedule inspections at different stages of the project to ensure quality standards are met.
    - **Third-party Audits:** Consider bringing in third-party experts to review the quality of work, especially for significant projects.
    - **Feedback Loop:** Use findings from inspections and audits to improve processes continually.
4. **Documentation:**
    - **Maintain Records:** Keep detailed records of all quality control measures, inspections, and tests.
    - **Review:** Regularly review these records to ensure compliance and identify areas for improvement.
5. **Final Inspection:**
    - **Checklist:** Use a comprehensive checklist to ensure all aspects of the project meet the required standards.
    - **Client Walk-through:** Before handing over the project, walk through it with the client to ensure their satisfaction and address any concerns.

Impact of Quality Control on a Contractor's Success and Reputation:

1. **Reputation Building:** Consistently delivering quality work enhances a contractor's reputation, making them a preferred choice for clients.
2. **Reduced Liabilities:** Quality control reduces the chances of structural failures or issues, thereby decreasing potential liabilities and legal challenges.
3. **Financial Benefits:** Fewer defects mean fewer reworks, leading to cost savings. Additionally, a good reputation can command better prices in the market.
4. **Client Trust:** A track record of quality work fosters trust, leading to repeat business and referrals.
5. **Regulatory Goodwill:** Consistent adherence to building codes and regulations can lead to smoother interactions with regulatory bodies.

In the context of the contractor's license exam, understanding the principles and practices of quality control is crucial. It reflects a contractor's commitment to excellence, safety, and client satisfaction, all of which are foundational to a successful career in construction.

Construction projects are complex endeavors, and various challenges can arise during their execution. Let's delve into some of the most common problems and the strategies to address them:

1. Scope Changes (Scope Creep):

- **Problem:** Changes or additions to a project's initial objectives can lead to increased costs, delays, and potential disputes.
- **Strategies:**
    - **Clear Documentation:** Ensure that the project's scope is well-defined and documented from the outset.
    - **Change Order Process:** Establish a clear process for handling change requests, including assessing the impact on time and budget.
    - **Regular Communication:** Maintain open lines of communication with clients and stakeholders to manage expectations.

2. Weather Disruptions:

- **Problem:** Adverse weather conditions can halt work, damage materials, and delay projects.
- **Strategies:**
    - **Weather Monitoring:** Use advanced weather forecasting tools to anticipate disruptions.
    - **Flexible Scheduling:** Build some flexibility into the project schedule to account for potential weather-related delays.
    - **Protective Measures:** Use tarps, temporary structures, or other means to protect work areas and materials from the elements.

3. Labor Issues:

- **Problem:** Strikes, high turnover, or a shortage of skilled labor can impede progress.
- **Strategies:**
    - **Fair Compensation:** Offer competitive wages and benefits to attract and retain workers.
    - **Training Programs:** Invest in training programs to upskill workers and reduce dependency on external skilled labor.
    - **Open Dialogue:** Maintain open communication with labor unions and workers to address grievances proactively.

4. Material Shortages:

- **Problem:** Delays in material deliveries or shortages can halt work and increase costs.
- **Strategies:**
    - **Early Procurement:** Order essential materials well in advance, especially if they have long lead times.

- **Alternative Suppliers:** Establish relationships with multiple suppliers to have backup options.
- **Inventory Management:** Use inventory management systems to track materials and anticipate shortages.

5. **Budget Overruns:**
   - **Problem:** Costs exceeding the budget can strain relationships with clients and result in financial losses.
   - **Strategies:**
     - **Detailed Estimation:** Ensure thorough cost estimation at the project's outset, accounting for potential contingencies.
     - **Regular Monitoring:** Continuously monitor expenses against the budget, flagging any potential overruns early.
     - **Transparent Communication:** Keep clients informed about any potential changes to the budget due to unforeseen circumstances.
6. **Communication Breakdowns:**
   - **Problem:** Misunderstandings or lack of communication can lead to mistakes, rework, and disputes.
   - **Strategies:**
     - **Regular Meetings:** Schedule regular project update meetings with all stakeholders.
     - **Collaboration Tools:** Use project management and collaboration tools to keep everyone updated.
     - **Clear Documentation:** Ensure all decisions, changes, and communications are well-documented.
7. **Safety Incidents:**
   - **Problem:** Accidents or safety violations can result in injuries, legal issues, and project delays.
   - **Strategies:**
     - **Safety Training:** Provide regular safety training to all workers.
     - **Safety Protocols:** Establish and enforce strict safety protocols on the construction site.
     - **Regular Inspections:** Conduct regular safety inspections to identify and rectify potential hazards.

For the contractor's license exam, understanding these challenges and their solutions is crucial. Effective problem-solving and risk management are hallmarks of a successful contractor, ensuring projects are delivered on time, within budget, and to the required quality standards.

Communication is the linchpin of successful construction project management. Let's delve into its role, strategies for ensuring clear communication, and the resultant impact:

Role of Communication in Construction Project Management:

1. **Clarifying Objectives:** Clear communication ensures that everyone understands the project's goals, scope, and deliverables.

2. **Decision Making:** Effective communication facilitates informed decision-making by ensuring that all stakeholders have the necessary information.
3. **Risk Management:** By communicating potential risks and challenges, teams can proactively address issues before they escalate.
4. **Coordination:** With multiple teams and subcontractors involved, communication is vital to coordinate tasks and ensure work progresses smoothly.
5. **Building Trust:** Transparent communication fosters trust between stakeholders, especially when addressing challenges or changes.

Ensuring Clear Communication with Stakeholders:

1. **Regular Updates:** Schedule regular status meetings and updates with clients, subcontractors, and other stakeholders.
2. **Use of Technology:** Employ project management software and communication tools to keep everyone updated in real-time.
3. **Clear Documentation:** Ensure all project details, changes, decisions, and communications are documented. This provides a reference point and reduces misunderstandings.
4. **Open Door Policy:** Encourage team members, subcontractors, and stakeholders to voice concerns, ask questions, and provide feedback.
5. **Feedback Mechanisms:** Implement mechanisms to gather feedback from stakeholders, helping to identify areas for improvement.
6. **Tailored Communication:** Recognize that different stakeholders require different types of information. Tailor communication to suit the audience – for instance, clients might want high-level updates, while subcontractors need detailed task-specific instructions.
7. **Conflict Resolution:** Establish clear procedures for resolving disputes or misunderstandings to ensure they're addressed promptly and fairly.

Impact of Effective Communication:

1. **Enhanced Project Outcomes:** Clear communication ensures that tasks are executed correctly the first time, reducing costly mistakes and rework.
2. **Timely Project Completion:** With everyone on the same page, delays due to misunderstandings or lack of information are minimized.
3. **Budget Adherence:** Effective communication around budget changes or potential overruns ensures financial transparency and can prevent unexpected costs.
4. **Increased Client Satisfaction:** Clients who are kept in the loop and feel their concerns are addressed are more likely to be satisfied with the project outcome.
5. **Strengthened Reputation:** Effective communication leads to successful projects, enhancing the contractor's reputation and increasing opportunities for future projects.
6. **Reduced Conflicts:** Transparent and proactive communication reduces the potential for disputes or conflicts among stakeholders.
7. **Compliance and Safety:** Clear communication with regulatory bodies ensures that all work complies with relevant codes and regulations. Similarly, effective communication of safety protocols reduces accidents.

For the contractor's license exam, understanding the pivotal role of communication in construction project management is essential. A contractor who prioritizes effective communication is more likely to deliver successful projects and foster strong, lasting relationships with clients and other stakeholders.

Safety Section:

Introduction to OSHA Standards:

The Occupational Safety and Health Administration (OSHA) was established in 1970 under the Occupational Safety and Health Act signed by President Richard Nixon. The primary purpose of OSHA is to ensure that every American worker has safe and healthful working conditions. This was a pivotal moment in U.S. labor history, as before OSHA's inception, countless workers faced hazardous conditions without any standardized protection.

In the construction industry, where the risks are inherently high due to the nature of the work, OSHA's role is paramount. The agency's standards have significantly reduced the number of accidents, injuries, and fatalities on construction sites. For instance, before OSHA, it was not uncommon for construction workers to operate without hard hats, safety harnesses, or proper scaffolding. Today, such practices are not only rare but also illegal and can result in hefty fines. OSHA's influence in the construction sector has led to the development and implementation of rigorous safety protocols. These protocols cover everything from the proper storage of materials to the safe operation of heavy machinery. The standards have made safety a core value in the construction industry, with many companies going beyond OSHA's requirements to ensure the well-being of their workers.

OSHA Standards for Specific Trades:

Each trade within the construction industry has its own set of challenges and risks. Recognizing this, OSHA has developed standards tailored to address the specific needs and potential hazards of each trade.

- **Electrical:** OSHA's standards for electrical workers revolve around preventing electrocutions and electrical burns. This includes guidelines on the proper use of electrical equipment, grounding practices, and the use of personal protective equipment (PPE) like insulating gloves and sleeves.
- **Plumbing:** For plumbers, OSHA standards emphasize preventing falls, trench collapses, and exposure to harmful substances. There are also guidelines on safely working with pressurized systems and handling tools and materials.
- **Masonry:** Masonry work can expose workers to risks like falling from heights, being struck by falling objects, and inhaling silica dust. OSHA standards for masonry address these risks with guidelines on scaffold use, mortar mixing, and the use of respirators when cutting or breaking materials that produce silica dust.

- **Carpentry:** Carpenters often work with power tools and at heights. OSHA's standards for carpentry focus on safely operating tools, preventing falls from elevated surfaces, and protecting against repetitive stress injuries.

In essence, OSHA's trade-specific standards are designed to address the unique challenges of each trade, ensuring that workers are protected from the most common and most severe risks they face. Over the years, these standards have undoubtedly saved countless lives and prevented innumerable injuries, making construction sites safer places to work.

Safety Equipment and OSHA Compliance:

Safety equipment, often referred to as Personal Protective Equipment (PPE), is a cornerstone of OSHA's safety regulations. The type of PPE required varies based on the specific job task and potential hazards:

- **Hard Hats:** Mandatory on almost all construction sites to protect workers from falling objects, electrical shocks, and bumps.
- **Safety Glasses/Goggles:** Essential for tasks like welding, cutting, nailing, or any operation that might send debris flying.
- **Hearing Protection:** Required in noisy environments, such as those involving heavy machinery or power tools.
- **Respirators:** Used when working in environments with dust, toxic fumes, or other harmful airborne particles. This is especially relevant for tasks like sanding or working with chemicals.
- **Safety Harnesses:** Crucial for workers operating at heights, preventing falls.
- **Protective Footwear:** Steel-toed boots or other protective footwear is essential on sites where there's a risk of foot injuries from falling objects or punctures.

Non-compliance with PPE regulations can result in hefty fines. The exact penalty varies based on the severity of the violation and whether it's a repeat offense. In some cases, OSHA can halt work on a site until violations are addressed.

OSHA's Role in Incident Reporting and Investigations:

When a severe workplace incident occurs, especially one that results in hospitalization, amputation, or loss of an eye, employers are required to report it to OSHA within 24 hours. Fatalities must be reported within 8 hours.

Upon receiving a report, OSHA may conduct an investigation. This process involves:

- **Site Inspection:** OSHA representatives will visit the site to inspect the incident area, interview workers, and review any relevant documentation.
- **Violation Identification:** If OSHA identifies any violations during their inspection, they'll inform the employer.
- **Penalties:** Based on the severity and nature of the violation, OSHA can impose fines. In extreme cases, criminal charges can be filed.

Staying Updated with OSHA Standards:

OSHA standards are not static; they evolve based on new research, technology, and industry needs. For contractors, staying updated is not just about compliance but ensuring the safety of their workforce.
OSHA offers a variety of resources:

- **Training Programs:** OSHA's Outreach Training Program provides training on the recognition, avoidance, abatement, and prevention of safety and health hazards.
- **Online Resources:** OSHA's official website is a treasure trove of information, from detailed standards documentation to educational videos.
- **Consultation Services:** For small and medium-sized businesses, OSHA offers free and confidential safety and health advice.

Contractors should also consider joining industry associations, attending seminars, and subscribing to industry publications to stay informed about the latest in safety standards and best practices.

Job Site Safety:

Maintaining a Safe Construction Job Site

A construction site, by its very nature, is rife with potential hazards. The foundational principle of job site safety is the proactive identification and mitigation of these hazards to prevent accidents and injuries. This is achieved through a combination of planning, training, equipment, and vigilance.

Regular Site Inspections:
Regular site inspections are a proactive approach to identify potential hazards before they become incidents. These inspections involve a thorough walkthrough of the site, checking equipment, evaluating work practices, and ensuring that safety protocols are being followed. By catching and addressing issues early, contractors can prevent many accidents.

Safety Meetings:
Often conducted at the start of a workday or before a new phase of construction begins, safety meetings serve to inform and remind workers about specific hazards associated with their tasks. These meetings provide an opportunity for workers to ask questions, share concerns, and be updated on any changes in procedures or new potential risks.

Hazard Assessments:
Before starting a new task or phase of construction, a hazard assessment identifies potential dangers associated with that task. This assessment considers everything from the tools and equipment needed to environmental conditions. Once identified, measures are put in place to mitigate those hazards.

Clear Signage:

Clear and visible signage plays a crucial role in safety. Signs can indicate hazardous zones, provide warnings about specific dangers (like overhead loads), or give directions (like emergency exit routes). Proper signage ensures that everyone on site, from seasoned workers to first-time visitors, is aware of and can avoid potential hazards.

Designated Walkways:
With so many moving parts and active work zones, it's essential to have designated walkways on a construction site. These walkways keep unauthorized personnel out of hazardous areas and provide a safe path for workers and visitors. They should be kept clear of obstructions and marked with clear signage.

Proper Site Housekeeping:
A tidy site is a safer site. Proper site housekeeping involves the regular removal of waste materials, ensuring tools and equipment are properly stored when not in use, and clearing walkways of debris. This not only reduces tripping hazards but also ensures that tools and materials are where they're expected to be, reducing the risk of accidents due to unexpected obstacles.

Practical Application:
Consider a multi-story building project. Before crane operations begin to hoist steel beams to upper levels, a safety meeting might be called to discuss the day's lifting plan. Clear signage would be placed to warn of overhead loads, and designated walkways ensure workers and visitors are safely routed around the crane's operating zone. Regular inspections would ensure the crane is in good working order, and hazard assessments might be conducted to account for factors like wind conditions. Proper site housekeeping ensures that once beams are in place, no leftover materials or tools are left at height, where they could pose a falling risk.

In essence, maintaining a safe construction job site is a continuous, multifaceted process. It requires the commitment of everyone involved, from site managers to individual laborers, to ensure that safety protocols are consistently followed and adapted as conditions change.

Significance of Personal Protective Equipment (PPE) in Construction
In the construction industry, where hazards are abundant and ever-present, PPE acts as the last line of defense against injuries. While engineering controls, administrative controls, and safe work practices are essential, PPE provides direct protection to workers from specific hazards when other methods are not feasible or do not provide sufficient protection.
Types of PPE and Their Applications:

1. **Hard Hats:** Designed to protect the head from falling objects, electrical hazards, and bumps. They are essential in areas where overhead work is being conducted or where there's a risk of head impact.
2. **Safety Glasses and Goggles:** These shield the eyes from flying debris, dust, chemicals, and other hazards. Depending on the task, workers might use basic safety glasses,

goggles that provide a seal around the eyes, or face shields for tasks with a high potential for splashing, like cutting or grinding.

3. **Respirators:** These protect workers from inhaling harmful substances. There are two main types:
    - *Particulate respirators* that filter out dust and particles.
    - *Gas/vapor respirators* that protect against harmful chemicals. Some tasks, like asbestos removal or working in confined spaces, might require specialized respirators.
4. **Protective Footwear:** Safety boots or shoes are designed to protect feet from punctures, crushing hazards, and slips, trips, and falls. Many have steel or composite toes and may be designed to be electrically insulating or resistant to chemicals.
5. **Hearing Protection:** Earplugs and earmuffs shield workers from loud noises, which are common on construction sites and can lead to long-term hearing loss.
6. **High-Visibility Clothing:** Brightly colored vests or jackets ensure workers are visible, especially crucial near moving vehicles or in low-light conditions.
7. **Gloves:** Depending on the task, workers might need gloves that protect against cuts, chemicals, electrical hazards, or cold temperatures.
8. **Fall Protection:** For work at heights, harnesses, lanyards, and other fall arrest systems are vital to prevent and mitigate the effects of falls.

Proper Fit, Maintenance, and Storage of PPE:

- **Fit:** A proper fit is crucial. Ill-fitting PPE can be uncomfortable, leading to reduced compliance among workers. More critically, it might not provide the intended protection. For instance, safety glasses that don't fit well might allow debris to enter from the sides, and a loose-fitting respirator might not provide an adequate seal, allowing harmful substances to be inhaled.
- **Maintenance:** Regular inspection and maintenance of PPE are essential. Damaged or worn-out PPE can compromise its protective qualities. For instance, a hard hat that's been struck hard might have compromised integrity, even if the damage isn't visible.
- **Storage:** Proper storage ensures PPE retains its protective qualities. For example, respirators stored in a dusty environment might become contaminated, and safety glasses stored without a protective case can become scratched, reducing visibility.

Practical Application: Consider a worker involved in welding operations. They would require safety goggles or a face shield with the appropriate shade to protect against the intense light and UV radiation. Leather gloves would protect their hands from sparks, and a respirator might be necessary to protect against fumes. If this worker's goggles were scratched or didn't fit well, they might be exposed to harmful UV radiation, leading to "welder's flash" or arc eye, a painful condition.

In essence, PPE is a critical component of construction safety. While the goal is always to eliminate hazards at the source or reduce exposure, PPE ensures that when workers are exposed, they have the best possible protection. Proper selection, use, and care of PPE are all essential components of a comprehensive safety program.

Best Practices for Safely Using, Storing, and Maintaining Tools and Heavy Equipment:

1. **Training Sessions:** Before any worker is allowed to handle tools or operate heavy equipment, they should undergo thorough training sessions. This ensures they understand the tool's or equipment's operation, potential hazards, and safety protocols. For instance, an operator of a crane should be well-versed in its load limits, control mechanisms, and emergency shutdown procedures.
2. **Equipment Inspections:** Regular inspections are crucial. Before using any tool or equipment, a visual and functional check should be conducted to identify any signs of wear, damage, or malfunction. For example, a frayed electrical cord on a power tool can pose an electrocution risk.
3. **Proper Tool Ergonomics:** Tools should be designed and used in a manner that minimizes stress on the user's body. Ergonomically designed tools reduce the risk of musculoskeletal disorders. For instance, a hammer with a shock-absorbing handle can reduce strain on the wrist and elbow.
4. **Using the Right Tool for the Job:** This cannot be emphasized enough. Using a tool for a purpose other than its intended use not only risks damaging the tool but also increases the risk of injury. For example, using a screwdriver as a chisel can cause it to break or slip, leading to potential injuries.
5. **Risks of Improvisation:** Improvising, like using a wrench instead of the correct socket size, can lead to tool slippage, bolt stripping, or even personal injury. It's essential to wait and use the correct tool rather than improvising with what's on hand.
6. **Safe Storage:** Tools should be stored in a designated place, such as a toolbox or storage shed. This prevents tripping hazards and also ensures tools are protected from the elements, which can cause rust or degradation. Heavy equipment should be parked on level ground and secured to prevent unauthorized use.
7. **Maintenance:** Regular maintenance, as per the manufacturer's guidelines, ensures tools and equipment function correctly and safely. This includes tasks like sharpening blades, lubricating moving parts, and replacing worn components.
8. **Safety Features:** Many modern tools and pieces of equipment come with built-in safety features, such as guards, automatic shut-offs, and anti-vibration mechanisms. Workers should be trained never to bypass these features and to report immediately if they're not functioning correctly.

Practical Application: Consider a table saw, a common tool on many construction sites. Before use, the saw should be inspected to ensure the blade is sharp, the guard is in place, and the emergency shut-off is functional. The operator should wear safety glasses, ear protection, and avoid loose clothing. They should be trained to feed wood at a steady rate, never reaching over the blade. If the saw were to jam, the worker should know to shut it off immediately and clear the obstruction safely.

In essence, tools and heavy equipment, while indispensable to the construction industry, come with inherent risks. Proper training, regular inspections, and a commitment to safety protocols are essential to minimizing these risks. For contractors, ensuring these practices are followed not only protects their workforce but also reduces downtime, liability, and potential regulatory penalties.

Common Hazards on Construction Sites:

1. **Falls:** This is one of the leading causes of fatalities in the construction industry. Falls can occur from elevated areas like roofs, ladders, and scaffolding or even on the same level due to tripping hazards.
    - **Mitigation Measures:** Use guardrails, safety nets, and fall arrest systems. Ensure ladders and scaffolds are secure and used correctly. Keep work areas free of clutter to prevent tripping.
2. **Electrocutions:** Contact with power lines, lack of proper grounding, and the use of faulty equipment can lead to electrocution.
    - **Mitigation Measures:** Always locate and identify utilities before starting work. Use ground-fault circuit interrupters. Regularly inspect electrical tools and equipment. Maintain a safe distance from overhead power lines.
3. **Struck-by-Object Incidents:** Workers can be injured by flying, falling, swinging, or rolling objects. This includes being hit by a vehicle, equipment, or even a tool dropped from above.
    - **Mitigation Measures:** Wear hard hats and high-visibility clothing. Use toe boards, screens, or debris nets to prevent objects from falling. Establish barricades and warning signs for areas where heavy equipment is operated.
4. **Caught-in/Between Accidents:** These involve workers being crushed, squeezed, compressed, or otherwise caught between two or more objects. Examples include trench collapses or getting caught in machinery.
    - **Mitigation Measures:** Always use protective systems for trenches like shoring, benching, or sloping. Ensure machinery has proper guards in place and that workers are trained not to wear loose clothing or jewelry that could get caught.

Proactive Measures:

1. **Regular Training:** Conduct regular safety training sessions, ensuring that all workers are familiar with potential hazards and know how to use equipment safely.
2. **Site Inspections:** Regularly inspect the job site to identify and rectify potential hazards.
3. **Safety Equipment:** Ensure that all necessary safety equipment, from PPE to machinery guards, is available, in good condition, and used consistently.
4. **Clear Communication:** Establish clear lines of communication for reporting hazards and potential risks.

Emergency Response and First Aid Preparedness:

1. **Emergency Response Plans:** Every construction site should have a clear and comprehensive emergency response plan. This plan should detail the steps to take in various emergency scenarios, from fires to chemical spills.
2. **First Aid Kits:** Sites should be equipped with easily accessible first aid kits tailored to the specific hazards of the job. Workers should be trained in basic first aid procedures.

3. **Emergency Contacts:** Maintain a list of emergency contacts, including the nearest hospital or medical facility, fire department, and other relevant agencies.
4. **Regular Drills:** Conduct regular emergency drills to ensure everyone knows what to do in case of an actual emergency.
5. **Incident Reporting:** Establish a system for reporting and documenting any incidents or near misses. This not only aids in addressing immediate concerns but also helps in identifying patterns and areas for improvement.

In essence, while construction sites inherently come with risks, a proactive approach to safety, combined with preparedness for emergencies, can significantly reduce the likelihood of accidents and ensure a swift and effective response when they do occur.

First aid, especially in the high-risk environment of construction sites, is the immediate and temporary care given to an individual who has suffered an injury or illness until professional medical assistance can be provided. The primary objectives of first aid are to preserve life, prevent the condition from worsening, and promote recovery.

Common Injuries on Construction Sites and First Aid Responses:

1. **Cuts and Lacerations:**
    - **First Aid:** Clean the wound with clean water to remove any debris. Apply gentle pressure with a sterile bandage or cloth to stop bleeding. Once bleeding is controlled, cover the wound with a sterile dressing.
    - **Note:** Deep cuts, especially from rusty objects, might require a tetanus shot or stitches.
2. **Burns:**
    - **First Aid:** For first-degree burns, cool the burn under cold running water for at least 10 minutes. Protect the burn from further injury by covering it with a non-stick bandage. For more severe burns, seek medical attention immediately.
    - **Note:** Chemical burns should be flushed with water immediately, and medical help should be sought.
3. **Fractures:**
    - **First Aid:** Immobilize the injured area using splints, boards, or rolled-up newspapers to keep the bone in place. Do not attempt to realign the bone. Elevate the area if possible and seek medical attention immediately.
4. **Electrical Shocks:**
    - **First Aid:** Do not touch the person if they're still in contact with the electrical source. Turn off the power source or use a non-conductive material, like wood, to separate them from the source. Once safe, check their breathing and pulse. Begin CPR if necessary and call for emergency medical help.

The DRSABCD Approach:

- **Danger:** Ensure the area is safe for both the rescuer and the casualty. On construction sites, this might mean turning off machinery, isolating electrical sources, or moving away from hazardous materials.

- **Response:** Check if the casualty is responsive by asking them questions or gently squeezing their shoulders.
- **Send for Help:** If the person is unresponsive, call for emergency services immediately or delegate someone else to do so.
- **Airway:** Check the casualty's airway to ensure it's clear. Remove any obstructions if necessary.
- **Breathing:** Check if the casualty is breathing. If they're not breathing normally, start CPR.
- **CPR (Cardiopulmonary Resuscitation):** Begin chest compressions. Place the heel of one hand on the center of the person's chest, place the other hand on top, and press down firmly and smoothly 30 times. Follow with two rescue breaths.
- **Defibrillation:** If an automated external defibrillator (AED) is available and the person is unresponsive and not breathing, apply the device and follow its instructions.

In the construction industry, where the risk of accidents is heightened due to the nature of the work, understanding and applying first aid principles can be the difference between life and death. Regular training sessions, drills, and having a well-stocked first aid kit accessible are crucial. Moreover, contractors should always ensure that at least several team members are certified in first aid and CPR on any given job site.

Responding to emergencies on a construction site requires a structured, systematic approach. The dynamic nature of construction sites, with their ever-changing environments and myriad potential hazards, necessitates a well-orchestrated response to ensure the safety of all involved.

Structured Process for Responding to Emergencies:

1. **Immediate Assessment:** As soon as an emergency arises, it's crucial to quickly assess the situation. Determine the nature and scale of the emergency. Is it a medical emergency, a fire, a chemical spill, or a structural collapse?
2. **Ensure Safety:** Before anything else, ensure your safety and then the safety of others. This might mean turning off machinery, isolating electrical sources, or moving away from hazardous materials.
3. **Alert and Delegate:**
    - **Raise the Alarm:** Use the site's emergency alarm system or shout to alert others.
    - **Delegate Roles:** Assign specific tasks to individuals based on their skills and proximity to resources. For instance, designate someone to call 911, another to administer first aid, and another to manage the crowd or direct an evacuation.
4. **Administer First Aid:** If someone is injured, trained personnel should administer first aid immediately while ensuring their safety. This might involve moving the injured person away from further harm, if possible, or providing CPR.
5. **Evacuation:** If the situation warrants, such as in the case of a fire or structural collapse:
    - Use established evacuation routes.
    - Ensure all workers move to the designated assembly point.
    - Account for all workers using a roll call or sign-in system.

- Never re-enter the site until it's declared safe by authorities.

6. **Manage Specific Hazards:**
    - **Fires:** Use fire extinguishers for small fires but prioritize evacuation for larger fires.
    - **Chemical Spills:** Isolate the area, prevent the spill from spreading, and use spill kits if available. Ensure that Material Safety Data Sheets (MSDS) for all chemicals on site are accessible.
    - **Structural Collapses:** Evacuate the area immediately. Ensure no one re-enters the danger zone until it's been assessed by structural engineers.
7. **Communication:** Once the immediate danger has passed, communicate the status of the emergency to all workers, inform higher management, and provide updates as necessary.
8. **Post-Emergency Review:** After every emergency, conduct a debriefing. Discuss what went well, what could have been done better, and how to improve future responses. Update emergency response plans based on lessons learned.

Key Considerations:

- **Quick Decision-Making:** In emergencies, seconds count. Trained personnel should be empowered to make decisions quickly without waiting for layers of approval.
- **Safety First:** Always prioritize the safety of individuals over equipment or structures. Material losses can be recovered; lost lives cannot.
- **Training:** Regular drills and training sessions ensure that all workers know their roles during emergencies, reducing chaos and ensuring a more efficient response.
- **Resources:** Ensure that emergency resources, such as first aid kits, fire extinguishers, and spill kits, are easily accessible and well-maintained.

In conclusion, while construction sites inherently come with risks, a well-prepared team with a clear emergency response plan can significantly mitigate these risks, ensuring the safety of all workers and minimizing potential damage.

Construction sites are rife with unique hazards, making specialized first aid knowledge indispensable. Let's delve into the proper responses to some common construction-related incidents and how contractors can ensure their teams are adequately prepared.

1. Falls from Heights:

- **Immediate Response:** Do not move the injured person unless they are in immediate danger. Immobilize the individual, especially the neck and spine.
- **First Aid:** Check for consciousness and breathing. If the person is not breathing, begin CPR. Control bleeding with direct pressure.
- **Prevention:** Ensure the use of fall protection equipment like harnesses and guardrails. Regularly inspect equipment and provide training on safe work at heights.

2. Machinery Entanglements:

- **Immediate Response:** Shut off the machinery immediately and lock out/tag out to ensure it doesn't restart. Do not attempt to extract the person without professional help unless it's life-threatening.

- **First Aid:** Once safely extricated, treat for shock and address any bleeding or broken bones.
- **Prevention:** Train workers on the safe operation of machinery, ensure machinery guards are in place, and promote the use of personal protective equipment (PPE) to prevent loose clothing or hair from getting caught.

3. Exposure to Hazardous Materials:

- **Immediate Response:** Identify the material involved. Remove the person from the exposure source and move them to fresh air if it's safe to do so.
- **First Aid:** For skin contact, flush the area with water for at least 15 minutes. If inhaled, ensure the person has access to fresh air. Always refer to the Material Safety Data Sheet (MSDS) for specific first aid measures.
- **Prevention:** Store hazardous materials properly, provide training on handling procedures, and ensure the availability of MSDS for all chemicals on site.

Equipping Teams with Necessary Skills and Resources:

1. **Regular Training:** Conduct frequent first aid training sessions tailored to construction hazards. This should include CPR, wound care, and specialized training for construction-specific risks.
2. **On-Site First Aid Kits:** Ensure that first aid kits are easily accessible throughout the site. These kits should be stocked with items specific to construction hazards, such as burn creams, eye wash stations, and splints.
3. **Emergency Plans:** Develop and regularly update emergency response plans. Ensure all workers are familiar with these plans and conduct regular drills.
4. **Safety Officers:** Appoint safety officers responsible for monitoring safety compliance, conducting regular inspections, and ensuring that workers are trained in first aid.
5. **Continuous Learning:** Regulations, best practices, and equipment evolve. Stay updated with the latest in first aid and safety protocols. OSHA and other regulatory bodies often offer resources and training programs.
6. **Feedback Mechanism:** Encourage workers to report near misses and provide feedback on safety protocols. This feedback can be invaluable in identifying potential hazards and improving safety measures.

In essence, while the nature of construction work brings inherent risks, a proactive approach to safety, combined with specialized first aid training, can significantly reduce the severity of injuries and improve overall site safety. For contractors, investing in safety not only protects their workforce but also enhances their reputation and reduces potential liabilities.

EXAM PREP SECTION:

Practice Test Question Section

Welcome to the Practice Test Question Section, a pivotal tool designed to bolster your understanding and readiness for the contractor's license exam. Each question in this section is meticulously crafted to mirror the type and caliber of questions you might encounter on the actual exam.

Now, you might notice something unique about our approach: the answer to each question, along with a detailed explanation, is provided immediately after the question itself. Why have we chosen this format?

1. **Immediate Feedback**: Instantly knowing the correct answer allows for immediate self-assessment. This real-time feedback mechanism can significantly enhance retention and understanding.
2. **Contextual Understanding**: By providing explanations right after the question, you're able to understand the context and rationale behind the correct answer. This not only aids in grasping the concept but also in avoiding similar mistakes in the future.
3. **Efficiency**: We value your time. Flipping to the back of the book repeatedly can be cumbersome and disrupt your flow. By placing answers and explanations adjacent to the questions, we aim to provide a seamless and efficient study experience.

Dive in, challenge yourself, and remember: every mistake is a learning opportunity. Let's begin!

1. A contractor is considering entering into a joint venture with another firm for a large project. Which of the following is NOT a typical advantage of a joint venture in the construction industry?

a. Pooling of resources and expertise.

b. Sharing of risks and liabilities.

c. Increased financial capacity for larger projects.

d. Simplified tax and regulatory compliance.

Answer: d. Simplified tax and regulatory compliance. Explanation: Joint ventures often introduce additional complexities in terms of tax and regulatory compliance due to the involvement of multiple entities. The primary advantages are pooling resources, sharing risks, and increasing financial capacity.

2. In the context of construction contracts, "liquidated damages" refer to:
a. Penalties for using too much liquid concrete.
b. Compensation for actual harm or loss suffered.
c. A predetermined amount to be paid for each day of delay.
d. Refunds given to clients for substandard work.

Answer: c. A predetermined amount to be paid for each day of delay.
Explanation: Liquidated damages are predetermined amounts specified in a contract, representing a fair estimate of damages in case of a breach, such as project delays.

3. Which of the following is NOT a primary purpose of a "mechanic's lien"?
a. To ensure contractors are paid for their work.
b. To place a legal claim on a property until a debt is settled.
c. To provide a warranty for the quality of work.
d. To give contractors a security interest in the property they've improved.

Answer: c. To provide a warranty for the quality of work.
Explanation: A mechanic's lien doesn't provide warranties. Its primary purpose is to ensure contractors are paid by granting them a security interest in the property they've worked on.

4. A contractor is reviewing a "time and materials" contract. This type of contract can be best described as:
a. A fixed-price agreement.
b. A contract where payment is based on actual costs plus a percentage or fixed fee.
c. A contract with a guaranteed maximum price.
d. An agreement based solely on project duration.

Answer: b. A contract where payment is based on actual costs plus a percentage or fixed fee.
Explanation: "Time and materials" contracts involve payment based on the actual costs of labor and materials, plus a markup (either a fixed fee or a percentage).

5. Which of the following best describes "indemnification" in a construction contract?
a. A clause ensuring timely payment to the contractor.
b. A provision where one party agrees to compensate another for certain liabilities.
c. A guarantee of work quality by the contractor.
d. A clause detailing the project's scope and specifications.

Answer: b. A provision where one party agrees to compensate another for certain liabilities.
Explanation: Indemnification refers to a party's promise to bear the financial burden for certain liabilities or losses that the other party might incur.

6. When a contractor is said to be "bonded," it means:
a. The contractor has a close relationship with the client.
b. The contractor has secured a loan for the project.
c. The contractor has purchased a bond ensuring project completion or compensation for non-completion.
d. The contractor has a partnership with other construction firms.

Answer: c. The contractor has purchased a bond ensuring project completion or compensation for non-completion.
Explanation: Being "bonded" means the contractor has a surety bond in place, which provides a financial guarantee that the contractor will fulfill their obligations.

7. In construction, "force majeure" clauses refer to:
a. The force exerted by heavy machinery.
b. The primary force behind a project's momentum.
c. Events beyond the control of parties, excusing them from liabilities.
d. The forceful removal of a contractor from a project.

Answer: c. Events beyond the control of parties, excusing them from liabilities.
Explanation: "Force majeure" clauses in contracts refer to unforeseen events (like natural disasters) that are beyond the control of the involved parties, potentially excusing them from fulfilling certain contractual obligations.

8. A "pay when paid" clause in a subcontractor's agreement means:
a. The subcontractor is paid only after the project's completion.
b. The subcontractor is paid immediately upon invoicing.
c. The subcontractor is paid when the prime contractor receives payment.
d. The subcontractor's payment is contingent on project approval.

Answer: c. The subcontractor is paid when the prime contractor receives payment.
Explanation: A "pay when paid" clause stipulates that a subcontractor will be paid once the primary or general contractor has been paid by the owner or client.

9. Which of the following is NOT typically covered under a contractor's general liability insurance?
a. Injuries to a third party at the job site.
b. Damage to the contractor's own equipment.
c. Damage to client property caused by the contractor's operations.
d. Claims of false advertising against the contractor.

Answer: b. Damage to the contractor's own equipment.
Explanation: General liability insurance typically covers third-party injuries, property damage, and some other claims like false advertising. Damage to the contractor's own equipment would usually be covered under a different type of insurance.

10. In the context of construction law, "statute of repose" refers to:
a. The time within which a contractor must respond to a project bid.
b. The period during which a building must be completed.
c. The maximum time after project completion during which legal action can be initiated for construction defects.
d. The time within which a contractor must repair any defects after project completion.

Answer: c. The maximum time after project completion during which legal action can be initiated for construction defects.
Explanation: A "statute of repose" sets a final deadline, after the completion of work, by which any legal action related to the project must be initiated. It protects contractors from indefinite liability.

11. In a typical slab-on-grade foundation, which component is primarily responsible for distributing the load of the structure to the soil below?
a. Vapor barrier
b. Reinforcing mesh
c. Concrete slab
d. Gravel or crushed stone base

Answer: c. Concrete slab
Explanation: The concrete slab in a slab-on-grade foundation directly distributes the load of the structure to the soil. While other components have their roles, it's the slab that primarily handles load distribution.

12. When considering the critical path method (CPM) in project management, which of the following best describes "float"?
a. The total time a task can be delayed without delaying the project.
b. The earliest start time for a task.
c. The total duration of the project.
d. The sequence of project activities.

Answer: a. The total time a task can be delayed without delaying the project.
Explanation: In CPM, "float" or "slack" refers to the amount of time a task can be delayed without causing a delay to subsequent tasks or the entire project.

13. Which of the following is a primary advantage of using post-tensioned concrete in construction?
a. Faster curing time
b. Enhanced resistance to tensile stresses
c. Reduced need for formwork
d. Lower concrete material cost

Answer: b. Enhanced resistance to tensile stresses
Explanation: Post-tensioning introduces compressive stress to counteract tensile stresses in the concrete, enhancing its tensile strength and reducing the risk of cracks.

14. In a building with a curtain wall system, what is the primary function of the mullions?
a. Provide structural support to the building.
b. Act as a decorative element.
c. Support and separate window panels.
d. Enhance thermal insulation.

Answer: c. Support and separate window panels.
Explanation: In curtain wall systems, mullions are the vertical or horizontal divisions that provide support for and separate the window, skylight, or curtain wall panels.

15. Which of the following is NOT a typical advantage of using a design-build delivery method in construction?
a. Faster project delivery
b. Single point of responsibility
c. Lower initial project cost
d. Enhanced design flexibility during construction

Answer: c. Lower initial project cost.
Explanation: While design-build can lead to cost savings over the project's lifecycle due to efficiencies and integrated teamwork, it doesn't necessarily guarantee a lower initial project cost compared to traditional delivery methods.

16. In green building practices, which of the following is a primary benefit of a green or "living" roof?
a. Enhanced structural support
b. Reduction in the heat island effect
c. Increased roof lifespan
d. Improved indoor air quality

Answer: b. Reduction in the heat island effect.
Explanation: Green roofs help mitigate the urban heat island effect by absorbing sunlight and providing insulation, thereby reducing the amount of heat reflected back into the environment.

17. When considering soil testing for construction, which test primarily determines the soil's ability to bear loads?
a. Atterberg limits test
b. Percolation test
c. Proctor compaction test
d. Plate bearing test

Answer: d. Plate bearing test.
Explanation: The plate bearing test is used to determine the bearing capacity of the soil, which indicates how much load the soil can support without excessive settlement.

18. In HVAC systems, which component is primarily responsible for removing heat and moisture from indoor air during the cooling cycle?
a. Furnace
b. Compressor
c. Evaporator coil
d. Condenser coil

Answer: c. Evaporator coil.
Explanation: The evaporator coil is where the refrigerant absorbs heat from the indoor air, causing it to evaporate and thereby cooling the air and removing moisture.

19. Which of the following is NOT a primary function of a building's facade or envelope?
a. Aesthetic appeal
b. Structural support
c. Thermal insulation
d. Weatherproofing

Answer: b. Structural support.
Explanation: While some facades can provide structural benefits, the primary functions of a building's facade or envelope are aesthetic appeal, thermal insulation, and weatherproofing.

20. In steel frame construction, which type of connection is designed to allow some degree of rotation between connected members?
a. Rigid connection
b. Pin connection
c. Fixed connection
d. Welded connection

Answer: b. Pin connection. Explanation: Pin connections in steel frame construction are designed to allow rotation between the connected members, making them partially restrained as opposed to fully restrained or fixed connections.

21. In the context of construction contracts, which of the following best describes the principle of "mutuality"?
a. Both parties must understand and agree to the terms.
b. Both parties must benefit financially from the contract.
c. Both parties must have prior experience in construction.
d. Both parties must have legal representation during negotiations.

Answer: a. Both parties must understand and agree to the terms. Explanation: The principle of mutuality means that both parties must have a mutual understanding and agreement on the terms for a contract to be valid.

22. A contractor fails to complete a project on time, citing unforeseen geological challenges as the reason. The client decides to terminate the contract. Which clause would most likely protect the contractor from penalties in this situation?
a. Indemnity clause
b. Force majeure clause
c. Liquidated damages clause
d. Exculpatory clause

Answer: b. Force majeure clause.
Explanation: A force majeure clause protects parties from unforeseen events beyond their control, which might prevent them from fulfilling contractual obligations.

23. Which of the following best describes a "unilateral contract" in the context of construction?
a. A contract where only one party makes a promise.
b. A contract that can be altered by one party without consent.
c. A contract where both parties exchange promises.
d. A contract that is signed by only one party.

Answer: a. A contract where only one party makes a promise.
Explanation: In a unilateral contract, one party makes a promise in exchange for the other party's performance. The party making the promise is obligated to fulfill it once the other party performs.

24. A construction contract states that any disputes will be resolved through arbitration rather than litigation. This is an example of:
a. An indemnity clause
b. An arbitration clause
c. A severability clause
d. A liquidated damages clause

Answer: b. An arbitration clause.
Explanation: An arbitration clause specifies that disputes will be resolved through arbitration rather than through the court system.

25. Which principle of contract law dictates that for a contract to be valid, there must be an exchange of value or something of worth between the parties?
a. Offer and acceptance
b. Legal purpose
c. Consideration
d. Capacity

Answer: c. Consideration.
Explanation: Consideration refers to the exchange of value in a contract, ensuring that both parties are giving and receiving something of worth.

26. In a construction contract, if a specific clause is found to be illegal or unenforceable, but the rest of the contract remains valid, this is due to which type of clause?
a. Arbitration clause
b. Severability clause
c. Indemnity clause
d. Exculpatory clause

Answer: b. Severability clause. Explanation: A severability clause ensures that if one part of a contract is deemed unenforceable, the rest of the contract remains in effect.

27. A contractor agrees to build a house for a client. The client agrees to pay upon completion. However, the contractor uses substandard materials not specified in the contract. The client refuses to pay. This situation is an example of:
a. Breach of contract
b. Force majeure
c. Unilateral mistake
d. Assignment of rights

Answer: a. Breach of contract. Explanation: Using materials not specified in the contract represents a breach of contract by the contractor, giving the client a potential reason to refuse payment.

28. Which of the following best describes "quantum meruit" in the context of construction contracts?
a. A contract based on specific quantities of materials.
b. A claim for the reasonable value of services rendered.
c. A contract that is void due to lack of consideration.
d. A claim for damages based on a breach of contract.

Answer: b. A claim for the reasonable value of services rendered.
Explanation: "Quantum meruit" is a Latin term meaning "as much as he has earned." It refers to a claim for a reasonable payment for services provided when no contract exists.

29. A construction company enters into a contract with a client. Later, the company transfers its contractual obligations to another company to complete the work. This transfer is known as:
a. Novation
b. Arbitration
c. Assignment
d. Estoppel

Answer: c. Assignment.
Explanation: Assignment refers to the transfer of rights or obligations under a contract from one party to another. The party making the transfer is the assignor, and the party receiving the rights or obligations is the assignee.

30. In the context of construction contracts, which principle dictates that both parties must genuinely agree to the terms without any form of deceit, duress, or undue influence?
a. Consideration
b. Genuine assent
c. Legal purpose
d. Offer and acceptance

Answer: b. Genuine assent.
Explanation: Genuine assent means that both parties have genuinely agreed to the terms of the contract without any form of deceit, duress, or undue influence affecting their decision.

31. Which of the following is NOT considered an essential element for a contract to be legally binding?
a. Consideration
b. Offer and acceptance
c. Notarization
d. Mutual assent

Answer: c. Notarization.
Explanation: While notarization can provide an additional layer of verification, it's not a fundamental requirement for most contracts to be legally binding. The essential elements include offer and acceptance, consideration, and mutual assent.

32. A contractor agrees to build a garage for a homeowner. The homeowner agrees to pay the contractor $20,000 upon completion. The $20,000 represents which essential element of a contract?
a. Offer
b. Acceptance
c. Consideration
d. Capacity

Answer: c. Consideration.
Explanation: Consideration refers to something of value exchanged between parties in a contract. In this case, the $20,000 is the consideration offered by the homeowner in exchange for the contractor's services.

33. Which element of a contract ensures that both parties enter into the agreement without any form of deceit, duress, or undue influence?
a. Legality of purpose
b. Genuine assent
c. Offer and acceptance
d. Consideration

Answer: b. Genuine assent.
Explanation: Genuine assent means that both parties have genuinely agreed to the terms of the contract without any form of deceit, duress, or undue influence affecting their decision.

34. In the context of contract law, which of the following best describes the "capacity" to contract?
a. The physical ability of a party to perform the work.
b. The legal competence of parties to enter into a contract.
c. The financial ability of a party to pay for services.
d. The scope of work outlined in the contract.

Answer: b. The legal competence of parties to enter into a contract.
Explanation: Capacity refers to the legal ability or competence of a party to enter into a contract. This includes considerations like age and mental competence.

35. A contractor sends a proposal to a client detailing the scope of work and the price. The client responds with a request for a slight change in the price. This response is best described as:
a. A counteroffer
b. An acceptance
c. A revocation
d. A unilateral contract

Answer: a. A counteroffer.
Explanation: When the client responds with a change to the terms presented in the original offer, it's considered a counteroffer. The original offer is effectively rejected, and the counteroffer must now be accepted for a contract to be formed.

36. Which of the following scenarios would likely render a contract void due to a lack of an essential element?
a. A contractor agrees to build a structure that violates local zoning laws.
b. A client agrees to pay a contractor upon completion of work.
c. A contractor provides a detailed scope of work in the contract.
d. A client provides a deposit as consideration.

Answer: a. A contractor agrees to build a structure that violates local zoning laws.
Explanation: For a contract to be legally binding, it must have a legal purpose. Agreeing to build a structure that violates local laws lacks legality of purpose, making the contract void.

37. A contractor and client discuss a potential project over the phone. They agree on terms but never put anything in writing. In the context of contract law, this type of agreement is known as:
a. A unilateral contract
b. An implied contract
c. An oral contract
d. A voidable contract

Answer: c. An oral contract.
Explanation: An oral contract is an agreement that's been discussed and agreed upon verbally without being put into written form. While oral contracts can be legally binding, they can be challenging to enforce due to a lack of concrete evidence.

38. For a contract to be legally binding, the offer presented must be:
a. Vague and general
b. Clear and definite
c. Open-ended
d. Subject to change

Answer: b. Clear and definite.
Explanation: For an offer to be valid and form a legally binding contract, it must be clear and definite, outlining specific terms and conditions.

39. If a party enters into a contract under duress or threat, which essential element of a contract is violated?
a. Consideration
b. Legality of purpose
c. Genuine assent
d. Offer and acceptance

Answer: c. Genuine assent.
Explanation: Genuine assent ensures that both parties agree to the terms without any form of deceit, duress, or undue influence. If a party is under duress, genuine assent is compromised.

40. A contractor agrees to complete a project for a client. However, the client does not promise anything in return. This agreement lacks which essential element of a contract?
a. Offer
b. Acceptance
c. Consideration
d. Capacity

Answer: c. Consideration.
Explanation: Consideration refers to something of value exchanged between parties in a contract. If one party doesn't promise anything in return, the contract lacks consideration and is not legally binding.

41. A contractor agrees to build a deck for a homeowner. The homeowner agrees to pay upon completion. Both parties sign a detailed written agreement outlining the terms. This type of contract is best described as:
a. Implied contract
b. Unilateral contract
c. Express contract
d. Oral contract

Answer: c. Express contract.
Explanation: An express contract is one where the terms are stated explicitly, either orally or in writing. In this case, the written agreement makes it an express contract.

42. In which type of contract does one party make a promise in exchange for the other party's performance, without a return promise?
a. Bilateral contract
b. Express contract
c. Unilateral contract
d. Implied contract

Answer: c. Unilateral contract.
Explanation: In a unilateral contract, one party makes a promise in exchange for the other party's performance. The party making the promise is obligated to fulfill it once the other party performs.

43. A contractor begins work on a homeowner's bathroom without a written agreement. The homeowner, seeing the work being done, does not object. This scenario is an example of:
a. Express contract
b. Unilateral contract
c. Implied contract
d. Bilateral contract

Answer: c. Implied contract.
Explanation: An implied contract is formed by the behavior of the parties involved. Even without a written or verbal agreement, the actions suggest an understanding or agreement.

44. Which of the following remedies for breach of contract requires the breaching party to perform their obligations as specified in the contract?
a. Liquidated damages
b. Specific performance
c. Cancellation
d. Consequential damages

Answer: b. Specific performance.
Explanation: Specific performance is a remedy that requires the breaching party to fulfill their obligations as outlined in the contract. It's often used when monetary damages are inadequate.

45. A contractor fails to complete a project on time, causing the client to lose potential rental income. The client may seek which type of damages?
a. Liquidated damages
b. Punitive damages
c. Consequential damages
d. Nominal damages

Answer: c. Consequential damages.
Explanation: Consequential damages, also known as special damages, arise from a specific result of the breach, such as lost rental income due to a delay in project completion.

46. In a construction contract, if a specific amount of money is agreed upon as compensation for breach or violation of the contract, this is known as:
a. Consequential damages
b. Liquidated damages
c. Punitive damages
d. Nominal damages

Answer: b. Liquidated damages.
Explanation: Liquidated damages are a predetermined amount of money that parties agree upon as compensation in the event of a breach.

47. Which type of contract involves mutual promises from both parties?
a. Unilateral contract
b. Implied contract
c. Express contract
d. Bilateral contract

Answer: d. Bilateral contract.
Explanation: In a bilateral contract, both parties exchange promises, meaning each party has both rights and obligations.

48. A contractor breaches a contract, and the client decides to terminate the agreement without seeking any damages. This action is an example of:
a. Specific performance
b. Cancellation
c. Liquidated damages
d. Consequential damages

Answer: b. Cancellation.
Explanation: Cancellation is a remedy where the non-breaching party decides to terminate the contract due to the breach.

49. In which scenario is an implied contract most likely formed?
a. A contractor provides a written estimate for a project.
b. A homeowner verbally agrees to a contractor's terms.
c. A contractor begins repairs without being asked but with the homeowner's knowledge.
d. A contractor and homeowner sign a detailed agreement.

Answer: c. A contractor begins repairs without being asked but with the homeowner's knowledge.
Explanation: An implied contract is formed by the behavior of the parties. If a contractor starts work and the homeowner, knowing this, doesn't object, an implied contract may be in place.

50. A contractor and client enter into an agreement. The contract states that if the contractor fails to complete the work on time, they will pay $200 for each day of delay. This type of provision is known as:
a. Consequential damages clause
b. Specific performance clause
c. Liquidated damages clause
d. Punitive damages clause

Answer: c. Liquidated damages clause.
Explanation: A liquidated damages clause specifies a predetermined amount of money that must be paid as compensation for breach or violation of the contract, such as delays.

51. A contractor claims they were under duress when they signed a construction contract. If proven, this claim can serve as:
a. An affirmation of the contract.
b. A defense to enforcement.
c. A reason for specific performance.
d. A cause for liquidated damages.

Answer: b. A defense to enforcement.
Explanation: Duress, if proven, can invalidate a contract, serving as a defense to its enforcement.

52. During contract negotiations, a contractor intentionally misrepresents the cost of materials to secure a higher project price. This is an example of:
a. Mistake
b. Duress
c. Fraud
d. Consideration

Answer: c. Fraud. Explanation: Intentional misrepresentation to induce another party into a contract is considered fraud, which can serve as a defense to the contract's enforcement.

53. Which of the following is NOT a valid reason for contract termination in a construction project?
a. Mutual agreement
b. Breach of contract
c. Completion of the project
d. Change of mind without cause

Answer: d. Change of mind without cause.
Explanation: While mutual agreement, breach of contract, and completion are valid reasons for termination, simply changing one's mind without a valid reason or cause does not justify contract termination.

54. In a real-world construction dispute, a contractor failed to use materials specified in the contract, leading to structural issues. The legal issue involved is likely:
a. Breach of contract
b. Fraudulent misrepresentation
c. Mutual mistake
d. Impossibility of performance

Answer: a. Breach of contract.
Explanation: Using materials other than those specified in the contract without agreement or justification is a breach of contract.

55. A construction company and a client are in a dispute over a delay in project completion. The contract specifies mediation as the first step in dispute resolution. This is an example of:
a. Liquidated damages clause
b. Specific performance clause
c. Arbitration clause
d. Mediation clause

Answer: d. Mediation clause.
Explanation: A mediation clause in a contract specifies that parties will attempt to resolve disputes through mediation before pursuing other legal remedies.

56. Which defense to contract enforcement arises when both parties have an incorrect belief about a material fact in the contract?
a. Unilateral mistake
b. Mutual mistake
c. Duress
d. Fraud

Answer: b. Mutual mistake.
Explanation: A mutual mistake occurs when both parties have an incorrect understanding or belief about a material fact essential to the contract.

57. In managing risks in construction projects, a clause that limits the amount a party can be held liable for is known as:
a. Indemnity clause
b. Limitation of liability clause
c. Force majeure clause
d. Liquidated damages clause

Answer: b. Limitation of liability clause.
Explanation: A limitation of liability clause sets a cap on the amount one party may have to pay the other in case of a breach or other issues.

58. A contractor is unable to complete a project because a new law bans the type of construction involved. This is an example of:
a. Breach of contract
b. Impossibility of performance
c. Mutual mistake
d. Fraud

Answer: b. Impossibility of performance.
Explanation: Impossibility of performance arises when unforeseen events, such as changes in law, make the contract's execution impossible.

59. During contract execution, a subcontractor demands a higher price than originally agreed upon, threatening to halt work if not paid the increased amount. This is an example of:
a. Mutual mistake
b. Fraud
c. Duress
d. Implied contract

Answer: c. Duress. Explanation: Threatening to halt work unless paid more than the agreed amount is a form of duress.

60. In a detailed analysis of a construction contract dispute, the primary cause was found to be ambiguous language in the contract leading to differing interpretations. This could have been avoided by:
a. Using standardized contract templates
b. Engaging in thorough contract review and negotiation
c. Relying on oral agreements
d. Avoiding technical jargon

Answer: b. Engaging in thorough contract review and negotiation.
Explanation: Ambiguities in contracts often arise from unclear language or terms. Thorough review and negotiation can help clarify and define terms, reducing the potential for disputes.

61. A contractor and client agree to change the type of materials used midway through a project. This agreement is an example of:
a. Breach of contract
b. Contract modification
c. Force majeure
d. Good faith negotiation

Answer: b. Contract modification.
Explanation: Contract modification refers to any change or alteration made to the original terms of a contract agreed upon by all parties.

62. During contract negotiations, a contractor intentionally hides certain costs to secure the project. This action is a violation of:
a. Good faith
b. Force majeure
c. Contract modification
d. Specific performance

Answer: a. Good faith.
Explanation: Good faith implies honesty and fairness in dealings. Intentionally hiding costs is a breach of this principle.

63. Which of the following is essential when drafting a construction contract to ensure clarity and avoid disputes?
a. Using complex legal jargon
b. Keeping the contract as brief as possible
c. Clearly defining terms and responsibilities
d. Relying on oral agreements for minor details

Answer: c. Clearly defining terms and responsibilities.
Explanation: Clearly defining terms, roles, and responsibilities in a contract ensures all parties understand their obligations, reducing potential disputes.

64. A construction project is halted due to an unexpected government-imposed lockdown. The contractor cannot be held liable for the delay due to:
a. Breach of contract
b. Good faith
c. Force majeure
d. Contract modification

Answer: c. Force majeure.
Explanation: Force majeure refers to unforeseen events outside the control of the contracting parties, like natural disasters or government actions, that prevent fulfilling contract obligations.

65. In contract execution, the concept of "good faith" primarily emphasizes:
a. Strictly adhering to the contract's written terms
b. Acting honestly and not taking advantage of the other party
c. Modifying the contract whenever necessary
d. Ensuring the contract is legally binding

Answer: b. Acting honestly and not taking advantage of the other party.
Explanation: Good faith implies honesty, fairness, and trustworthiness in contract execution and dealings.

66. Which of the following is NOT a common reason for contract modification in the construction industry?
a. Change in project scope
b. Unexpected site conditions
c. Mutual agreement to terminate the contract
d. Change in material prices

Answer: c. Mutual agreement to terminate the contract.
Explanation: While the first three options can lead to modifications, mutual agreement to terminate ends the contract rather than modifying it.

67. Legal and ethical considerations in contract negotiation in the construction industry are crucial because:
a. They ensure maximum profit for all parties
b. They protect against potential lawsuits and disputes
c. They allow for easier contract modifications
d. They ensure faster project completion

Answer: b. They protect against potential lawsuits and disputes.
Explanation: Adhering to legal and ethical standards in contract negotiation ensures clarity, fairness, and reduces the risk of disputes and legal actions.

68. When drafting a construction contract, which section outlines the steps to be taken in case of disagreements between parties?
a. Payment terms
b. Dispute resolution clause
c. Force majeure clause
d. Scope of work

Answer: b. Dispute resolution clause.
Explanation: The dispute resolution clause specifies the agreed-upon methods (like mediation or arbitration) for resolving disagreements.

69. In the context of the construction industry, acting in "good faith" during contract execution means:
a. Always choosing the cheapest materials
b. Completing the project ahead of schedule
c. Dealing honestly and transparently with all parties
d. Modifying the contract frequently

Answer: c. Dealing honestly and transparently with all parties.
Explanation: Good faith emphasizes honesty, transparency, and fairness in dealings, ensuring trust and smooth execution.

70. A contractor is unable to continue a project due to unexpected and severe weather conditions that pose safety risks. This scenario can be addressed in the contract under:
a. Good faith clause
b. Contract modification clause
c. Force majeure clause
d. Dispute resolution clause

Answer: c. Force majeure clause.
Explanation: The force majeure clause addresses unforeseen events, like extreme weather conditions, that are beyond the control of the contracting parties and may impact contract fulfillment.

71. Under the Fair Labor Standards Act (FLSA), which of the following is NOT a requirement for an employee to be considered exempt from overtime pay?
a. The employee must be paid on a salary basis.
b. The employee must earn at least $35,568 annually.
c. The employee must work less than 40 hours a week.
d. The employee must perform executive, administrative, or professional duties.

Answer: c. The employee must work less than 40 hours a week.
Explanation: The FLSA does not set a maximum limit on the number of hours an employee can work in a week to be considered exempt. The exemption is based on salary, salary level, and job duties.

72. Which federal act prohibits employers from discriminating against individuals based on race, color, religion, sex, or national origin?
a. Occupational Safety and Health Act (OSHA)
b. Fair Labor Standards Act (FLSA)
c. Equal Employment Opportunity Act (EEOA)
d. Americans with Disabilities Act (ADA)

Answer: c. Equal Employment Opportunity Act (EEOA).
Explanation: The EEOA enforces federal laws that prohibit employment discrimination.

73. A construction company is found to have not provided adequate safety training to its employees. This is a potential violation of:
a. Equal Employment Opportunity Act (EEOA)
b. Fair Labor Standards Act (FLSA)
c. Occupational Safety and Health Act (OSHA)
d. Americans with Disabilities Act (ADA)

Answer: c. Occupational Safety and Health Act (OSHA).
Explanation: OSHA sets and enforces standards to ensure safe working conditions.

74. Which of the following is NOT a protected category under the Equal Employment Opportunity Commission (EEOC) guidelines?
a. Age
b. Marital status
c. Race
d. Religion

Answer: b. Marital status.
Explanation: While many states have laws against marital status discrimination, it is not a federally protected category under EEOC guidelines.

75. A construction company refuses to hire a qualified individual because they use a wheelchair. This could be a violation of:
a. Occupational Safety and Health Act (OSHA)
b. Fair Labor Standards Act (FLSA)
c. Equal Employment Opportunity Act (EEOA)
d. Americans with Disabilities Act (ADA)

Answer: d. Americans with Disabilities Act (ADA).
Explanation: The ADA prohibits discrimination against individuals with disabilities in all areas of public life, including jobs.

76. Which act requires employers to provide eligible employees with up to 12 weeks of unpaid leave for certain medical and family reasons?
a. Family and Medical Leave Act (FMLA)
b. Fair Labor Standards Act (FLSA)
c. Occupational Safety and Health Act (OSHA)
d. Equal Employment Opportunity Act (EEOA)

Answer: a. Family and Medical Leave Act (FMLA).
Explanation: The FMLA entitles eligible employees to take unpaid, job-protected leave for specified family and medical reasons.

77. In the construction industry, which of the following is NOT a primary focus of the Occupational Safety and Health Act (OSHA)?
a. Ensuring safe working conditions
b. Setting and enforcing safety standards
c. Monitoring employee break times
d. Providing training and education in safety

Answer: c. Monitoring employee break times.
Explanation: While break times can be related to safety, OSHA's primary focus is on setting and enforcing standards to ensure safe working conditions.

78. A construction company pays female workers less than male workers for the same job role and responsibilities. This is a potential violation of:
a. Equal Pay Act
b. Occupational Safety and Health Act (OSHA)
c. Americans with Disabilities Act (ADA)
d. Family and Medical Leave Act (FMLA)

Answer: a. Equal Pay Act.
Explanation: The Equal Pay Act prohibits sex-based wage discrimination between men and women in the same establishment who perform jobs that require substantially equal skill, effort, and responsibility.

79. Which act prohibits employers from discriminating against employees or applicants because they are 40 years of age or older?
a. Age Discrimination in Employment Act (ADEA)
b. Fair Labor Standards Act (FLSA)
c. Occupational Safety and Health Act (OSHA)
d. Equal Employment Opportunity Act (EEOA)

Answer: a. Age Discrimination in Employment Act (ADEA).
Explanation: The ADEA prohibits employment discrimination against persons 40 years of age or older.

80. A construction worker who is a whistleblower reports safety violations to OSHA. Which act protects this worker from employer retaliation?
a. Whistleblower Protection Act
b. Occupational Safety and Health Act (OSHA)
c. Fair Labor Standards Act (FLSA)
d. Equal Employment Opportunity Act (EEOA)

Answer: a. Whistleblower Protection Act.
Explanation: The Whistleblower Protection Act protects federal employees and contractors who report agency misconduct.

81. Under the Fair Labor Standards Act (FLSA), which of the following is NOT a primary concern?
a. Minimum wage
b. Overtime pay
c. Child labor provisions
d. Safety equipment standards

Answer: d. Safety equipment standards.
Explanation: While the FLSA addresses wage, overtime, and child labor, safety equipment standards are primarily under the jurisdiction of OSHA.

82. A construction company that employs minors under the age of 16 during school hours is potentially in violation of:
a. Occupational Safety and Health Act (OSHA)
b. Family and Medical Leave Act (FMLA)
c. Fair Labor Standards Act (FLSA)
d. Age Discrimination in Employment Act (ADEA)

Answer: c. Fair Labor Standards Act (FLSA).
Explanation: The FLSA sets age restrictions and guidelines for the employment of minors, including restrictions on working during school hours.

83. The primary goal of the Occupational Safety and Health Act (OSHA) is to:
a. Ensure employees receive minimum wage
b. Guarantee job security for employees
c. Ensure a safe and healthful working environment
d. Ensure equal employment opportunities

Answer: c. Ensure a safe and healthful working environment. Explanation: OSHA's main objective is to ensure safe working conditions by setting and enforcing standards.

84. Which of the following is NOT a requirement under the Fair Labor Standards Act (FLSA)?
a. Employers must pay non-exempt employees time and a half for overtime.
b. Employers must provide health insurance to all employees.
c. Employers must pay at least the federal minimum wage.
d. Employers must keep accurate record of hours worked and wages paid.

Answer: b. Employers must provide health insurance to all employees. Explanation: While the FLSA mandates wage, overtime, and record-keeping standards, it does not require employers to provide health insurance.

85. A construction site that lacks proper fall protection measures is likely in violation of:
a. Fair Labor Standards Act (FLSA)
b. Family and Medical Leave Act (FMLA)
c. Occupational Safety and Health Act (OSHA)
d. Equal Employment Opportunity Act (EEOA)

Answer: c. Occupational Safety and Health Act (OSHA).
Explanation: OSHA sets and enforces standards to ensure safe working conditions, including fall protection measures in construction.

86. Under the FLSA, which of the following employees is most likely to be considered exempt from overtime pay?
a. A construction laborer
b. A site supervisor
c. A crane operator
d. A bricklayer

Answer: b. A site supervisor.
Explanation: Exemptions from FLSA overtime provisions typically apply to executive, administrative, and professional employees. A site supervisor would likely fall under this category.

87. A construction company is required to provide training on the hazards of asbestos exposure due to regulations set by:
a. Fair Labor Standards Act (FLSA)
b. Occupational Safety and Health Act (OSHA)
c. Family and Medical Leave Act (FMLA)
d. Age Discrimination in Employment Act (ADEA)

Answer: b. Occupational Safety and Health Act (OSHA).
Explanation: OSHA sets standards for training and protection against various workplace hazards, including asbestos.

88. Which act mandates that employers with 50 or more employees provide unpaid, job-protected leave for specified family and medical reasons?
a. Fair Labor Standards Act (FLSA)
b. Occupational Safety and Health Act (OSHA)
c. Family and Medical Leave Act (FMLA)
d. Equal Employment Opportunity Act (EEOA)

Answer: c. Family and Medical Leave Act (FMLA).
Explanation: The FMLA requires eligible employers to provide up to 12 weeks of unpaid, job-protected leave for certain family and medical reasons.

89. In the construction industry, which of the following would be a primary concern for OSHA?
a. Ensuring employees are paid on time
b. Ensuring safety equipment is up to standard
c. Ensuring employees receive paid vacation
d. Ensuring equal pay for equal work

Answer: b. Ensuring safety equipment is up to standard.
Explanation: OSHA's main focus is on setting and enforcing standards to ensure safe working conditions, including the use of proper safety equipment.

90. Under the FLSA, if a non-exempt construction worker works 50 hours in a week, how many hours of overtime pay are they entitled to?
a. 5 hours
b. 10 hours
c. 15 hours
d. 20 hours

Answer: b. 10 hours.
Explanation: The standard workweek is 40 hours. Any hours worked beyond that are considered overtime. In this case, 50 - 40 = 10 hours of overtime.

91. In the context of construction work, which of the following is NOT typically covered by Workers' Compensation?
a. Injuries sustained from a fall on-site
b. Illnesses related to asbestos exposure during a project
c. A car accident while commuting to the job site
d. Injuries from malfunctioning equipment

Answer: c. A car accident while commuting to the job site.
Explanation: Workers' Compensation generally covers injuries or illnesses that occur while performing job duties or as a direct result of job tasks. Commuting to and from work is typically not covered.

92. A construction company with how many employees is required to comply with the Family and Medical Leave Act (FMLA)?
a. 15 or more
b. 25 or more
c. 50 or more
d. 100 or more

Answer: c. 50 or more.
Explanation: The FMLA applies to employers with 50 or more employees in 20 or more workweeks in the current or preceding calendar year.

93. Which of the following is NOT a primary focus of Equal Employment Opportunity laws in the construction industry?
a. Preventing wage discrimination based on gender
b. Ensuring safe working conditions
c. Prohibiting discrimination based on race or religion
d. Providing equal opportunities for individuals with disabilities

Answer: b. Ensuring safe working conditions.
Explanation: While safety is crucial in the construction industry, it's primarily addressed by OSHA. Equal Employment Opportunity laws focus on preventing discrimination and ensuring equal opportunities.

94. A construction worker who is injured on the job and receives Workers' Compensation is typically barred from:
a. Returning to work
b. Suing the employer for the injury
c. Reporting the injury to OSHA
d. Seeking medical treatment

Answer: b. Suing the employer for the injury.
Explanation: One of the primary purposes of Workers' Compensation is to provide benefits to injured workers while also protecting employers from lawsuits related to those injuries.

95. Under the FMLA, eligible employees are entitled to up to how many weeks of unpaid, job-protected leave in a 12-month period for specified family and medical reasons?
a. 6 weeks
b. 10 weeks
c. 12 weeks
d. 16 weeks

Answer: c. 12 weeks.
Explanation: The FMLA entitles eligible employees to take up to 12 weeks of unpaid, job-protected leave in a 12-month period for certain family and medical reasons.

96. Which of the following is NOT a protected category under Equal Employment Opportunity laws?
a. Political affiliations
b. Race
c. Religion
d. Gender

Answer: a. Political affiliations.
Explanation: While many employers choose not to discriminate based on political affiliations, it's not a federally protected category under Equal Employment Opportunity laws.

97. If a construction worker is injured and unable to perform their regular job duties, Workers' Compensation may provide:
a. Job retraining
b. A promotion
c. A one-time bonus
d. An extended vacation

Answer: a. Job retraining.
Explanation: If an injury prevents a worker from returning to their previous job, Workers' Compensation may provide vocational rehabilitation or job retraining.

98. Under the FMLA, in order to be eligible for leave, an employee must have worked for their employer for at least:
a. 3 months
b. 6 months
c. 12 months
d. 24 months

Answer: c. 12 months. Explanation: To be eligible under the FMLA, an employee must have worked for their employer for at least 12 months and have worked at least 1,250 hours during the 12 months prior to the start of the FMLA leave.

99. In the context of Equal Employment Opportunity laws, which of the following best describes "affirmative action" in the construction industry?
a. A mandatory quota system for hiring minority workers
b. A voluntary program to increase opportunities for underrepresented groups
c. A system for reporting workplace injuries
d. A training program for new hires

Answer: b. A voluntary program to increase opportunities for underrepresented groups.
Explanation: Affirmative action refers to voluntary programs intended to increase employment opportunities for minority groups and redress past discrimination.

100. A construction company that fails to provide reasonable accommodations for an employee's religious practices may be in violation of:
a. Workers' Compensation laws
b. The FMLA
c. Equal Employment Opportunity laws
d. Wage and hour laws

Answer: c. Equal Employment Opportunity laws.
Explanation: Equal Employment Opportunity laws require employers to provide reasonable accommodations for an employee's religious beliefs and practices, unless doing so would cause undue hardship for the employer.

101. Which of the following best describes the primary purpose of the Davis-Bacon Act as it pertains to the construction industry?
a. To ensure equal employment opportunities for all workers.
b. To mandate safety standards on construction sites.
c. To require contractors to pay prevailing wages on federally funded projects.
d. To regulate the hours construction workers can work in a week.

Answer: c. To require contractors to pay prevailing wages on federally funded projects.
Explanation: The Davis-Bacon Act mandates that contractors and subcontractors must pay their workers no less than the prevailing wages in the area for corresponding work on similar projects.

102. In the construction industry, which of the following is NOT a typical component of performance management?
a. Annual salary reviews
b. Weekly safety training sessions
c. Regular feedback on job performance
d. Setting clear job expectations

Answer: b. Weekly safety training sessions.
Explanation: While safety training is crucial in the construction industry, it's not typically a component of performance management, which focuses on evaluating and improving employee performance.

103. A construction company that refuses to hire an applicant because of their age, when the applicant is over 40 years old, may be in violation of:
a. The Davis-Bacon Act
b. The Age Discrimination in Employment Act (ADEA)
c. The Occupational Safety and Health Act (OSHA)
d. The Fair Labor Standards Act (FLSA)

Answer: b. The Age Discrimination in Employment Act (ADEA).
Explanation: The ADEA prohibits employment discrimination against persons 40 years of age or older.

104. Which of the following is a potential consequence for a construction company found in violation of wage laws?
a. Mandatory safety training for all employees
b. Revocation of the company's business license
c. Payment of back wages to affected employees
d. Mandatory diversity training for hiring managers

Answer: c. Payment of back wages to affected employees. Explanation: Companies found in violation of wage laws are often required to pay back wages to employees who were underpaid.

105. In the context of hiring practices in the construction industry, which of the following is NOT considered a best practice?
a. Using a standardized interview process for all candidates
b. Hiring based solely on a candidate's physical strength
c. Checking references and past employment history
d. Offering onboarding and training for new hires

Answer: b. Hiring based solely on a candidate's physical strength. Explanation: While physical capabilities can be important for certain roles in construction, hiring based solely on physical strength can lead to discriminatory practices and overlook other essential skills and qualifications.

106. A construction company that consistently pays female workers less than male workers for the same job may be in violation of:
a. The Equal Pay Act
b. The Davis-Bacon Act
c. The Family and Medical Leave Act (FMLA)
d. The Employee Retirement Income Security Act (ERISA)

Answer: a. The Equal Pay Act.
Explanation: The Equal Pay Act prohibits wage discrimination between men and women in the same establishment who perform jobs that require substantially equal skill, effort, and responsibility.

107. In the construction industry, which of the following is a common reason for termination?
a. Refusal to attend mandatory diversity training
b. Consistent tardiness or absenteeism
c. Requesting time off under the FMLA
d. Reporting safety violations to OSHA

Answer: b. Consistent tardiness or absenteeism.
Explanation: Regular attendance is crucial in the construction industry, and consistent tardiness or absenteeism can be grounds for termination.

108. Which of the following laws prohibits discrimination in hiring based on race, color, religion, sex, or national origin in the construction industry?
a. The Davis-Bacon Act
b. The Fair Labor Standards Act (FLSA)
c. The Civil Rights Act (Title VII)
d. The Age Discrimination in Employment Act (ADEA)

Answer: c. The Civil Rights Act (Title VII).
Explanation: Title VII of the Civil Rights Act prohibits employment discrimination based on race, color, religion, sex, or national origin.

109. In terms of contract termination in the construction industry, which of the following is NOT a typical ground for termination for cause?
a. Failure to make timely payments
b. Minor delays in project completion
c. Persistent failure to follow safety standards
d. Use of substandard materials

Answer: b. Minor delays in project completion.
Explanation: While timely completion is important, minor delays, unless specified in the contract, are not typically grounds for termination for cause. Persistent or significant delays, however, might be.

110. Which of the following is a primary consideration when drafting a construction contract to manage risks?
a. Ensuring flexible work hours for all employees
b. Clearly defining the scope of work and responsibilities
c. Mandating weekly team-building activities
d. Setting a fixed salary for all workers regardless of role

Answer: b. Clearly defining the scope of work and responsibilities.
Explanation: Clearly defining the scope of work, responsibilities, and deliverables in a construction contract is essential to manage risks and set clear expectations for all parties involved.

111. How might labor laws directly influence the contractor-client relationship in a construction project?
a. By dictating the design aesthetics of the project.
b. By determining the materials used in the project.
c. By influencing the wage rates and working conditions.
d. By deciding the location of the construction site.

Answer: c. By influencing the wage rates and working conditions.
Explanation: Labor laws often dictate wage standards, working hours, and conditions, which can influence project costs and timelines, thereby affecting the contractor-client relationship.

112. Which of the following is a primary reason construction companies might prefer to avoid unionized labor?
a. Unionized labor often requires specialized equipment.
b. Unionized labor typically demands higher wages and benefits.
c. Union workers are less skilled than non-union workers.
d. Union contracts usually have shorter work hours.

Answer: b. Unionized labor typically demands higher wages and benefits.
Explanation: One of the primary objectives of labor unions is to secure better wages, benefits, and working conditions for their members, which can increase project costs.

113. In the context of labor unions in the construction industry, what is a "closed shop"?
a. A construction site that is closed to the public.
b. A company that only hires union members.
c. A company that refuses to work with subcontractors.
d. A union that does not allow new members.

Answer: b. A company that only hires union members.
Explanation: A "closed shop" refers to a workplace where the employer agrees to hire only union members.

114. How can a construction company ensure compliance with employment regulations?
a. By hiring only experienced workers.
b. By conducting regular training and audits.
c. By avoiding unionized labor.
d. By completing projects ahead of schedule.

Answer: b. By conducting regular training and audits.
Explanation: Regular training ensures that employees are aware of regulations, and audits help verify that practices comply with those regulations.

115. Which of the following is NOT a typical role of labor unions in the construction industry?
a. Negotiating wage rates and benefits.
b. Designing construction projects.
c. Advocating for safer working conditions.
d. Organizing strikes or work stoppages.

Answer: b. Designing construction projects. Explanation: Labor unions focus on worker rights, wages, and conditions, not on the design aspects of construction projects.

116. A construction company that consistently fails to pay overtime might be in violation of which act?
a. The Davis-Bacon Act
b. The Fair Labor Standards Act (FLSA)
c. The Occupational Safety and Health Act (OSHA)
d. The National Labor Relations Act (NLRA)

Answer: b. The Fair Labor Standards Act (FLSA). Explanation: The FLSA establishes minimum wage, overtime pay, and other employment standards.

117. Which of the following is a potential consequence for a construction company that interferes with its employees' rights to form a union?
a. Immediate shutdown of the construction site.
b. Revocation of the company's business license.
c. Penalties or sanctions from the National Labor Relations Board (NLRB).
d. Mandatory diversity training for all employees.

Answer: c. Penalties or sanctions from the National Labor Relations Board (NLRB).
Explanation: The NLRB enforces the National Labor Relations Act, which protects employees' rights to organize and to bargain collectively.

118. In the context of labor laws, what is the primary purpose of a collective bargaining agreement in the construction industry?
a. To outline the design specifications of a project.
b. To set the terms and conditions of employment.
c. To determine the location of the construction site.
d. To establish the company's annual budget.

Answer: b. To set the terms and conditions of employment.
Explanation: A collective bargaining agreement is a labor contract between an employer and one or more unions, setting the terms of employment.

119. Which of the following best describes the concept of "right-to-work" as it pertains to the construction industry?
a. The right of a worker to be paid for all hours worked.
b. The right of a worker to join or not join a union without facing coercion.
c. The right of a worker to receive training on safety protocols.
d. The right of a worker to refuse any task they deem unsafe.

Answer: b. The right of a worker to join or not join a union without facing coercion.
Explanation: "Right-to-work" laws prohibit union security agreements, or agreements between employers and labor unions, that govern the extent to which an established union can require employees' membership, payment of union dues, or fees as a condition of employment.

120. A construction company that discriminates against an employee because they filed a complaint about unsafe working conditions might be in violation of:
a. The Davis-Bacon Act
b. The Fair Labor Standards Act (FLSA)
c. The Occupational Safety and Health Act (OSHA)
d. The Equal Pay Act

Answer: c. The Occupational Safety and Health Act (OSHA).
Explanation: OSHA prohibits employers from retaliating against employees for exercising their rights under the Act, including the right to raise health and safety concerns.

121. Which of the following is a potential legal consequence for construction companies that violate wage and hour regulations?
a. Revocation of construction permits.
b. Mandatory safety training for all employees.
c. Fines and back pay for affected employees.
d. Immediate termination of company executives.

Answer: c. Fines and back pay for affected employees.
Explanation: Violating wage and hour regulations can result in fines and the requirement to pay back wages to affected employees.

122. In the context of employment regulations, how might subcontractors differ from regular employees in the construction industry?
a. Subcontractors are always paid a salary, while employees are paid hourly.
b. Subcontractors can't be required to work overtime.
c. Subcontractors typically provide their own tools and materials.
d. Subcontractors are always part of a union.

Answer: c. Subcontractors typically provide their own tools and materials.
Explanation: One distinction between subcontractors and regular employees is that subcontractors often supply their own tools and materials for a job.

123. A construction company that classifies workers as independent contractors to avoid paying benefits might face legal consequences under which act?
a. The Davis-Bacon Act
b. The Fair Labor Standards Act (FLSA)
c. The Occupational Safety and Health Act (OSHA)
d. The Employee Retirement Income Security Act (ERISA)

Answer: b. The Fair Labor Standards Act (FLSA).
Explanation: Misclassifying employees as independent contractors to avoid wage and benefit obligations can lead to violations under the FLSA.

124. Which of the following is NOT a typical characteristic of an independent contractor in the construction industry?
a. Receives a W-2 form at the end of the year.
b. Has control over how the work is performed.
c. Provides their own tools and equipment.
d. Is hired for a specific task or project.

Answer: a. Receives a W-2 form at the end of the year.
Explanation: Independent contractors typically receive a 1099 form, not a W-2, which is for employees.

125. In a recent legal dispute, a construction company faced penalties for not providing safety equipment to its workers. This is most likely a violation of which act?
a. The Davis-Bacon Act
b. The Fair Labor Standards Act (FLSA)
c. The Occupational Safety and Health Act (OSHA)
d. The National Labor Relations Act (NLRA)

Answer: c. The Occupational Safety and Health Act (OSHA).
Explanation: OSHA sets and enforces standards to ensure safe working conditions, including the provision of necessary safety equipment.

126. Which of the following is a potential consequence for construction companies that discriminate against employees based on age?
a. Revocation of construction permits.
b. Penalties under the Age Discrimination in Employment Act (ADEA).
c. Immediate shutdown of the construction site.
d. Mandatory diversity training for all subcontractors.

Answer: b. Penalties under the Age Discrimination in Employment Act (ADEA).
Explanation: The ADEA prohibits employment discrimination against persons 40 years of age or older.

127. A construction company that fails to provide reasonable accommodations for an employee with a disability might be in violation of:
a. The Davis-Bacon Act
b. The Americans with Disabilities Act (ADA)
c. The Occupational Safety and Health Act (OSHA)
d. The Employee Retirement Income Security Act (ERISA)

Answer: b. The Americans with Disabilities Act (ADA).
Explanation: The ADA requires employers to provide reasonable accommodations for employees with disabilities.

128. Which of the following is a primary reason construction companies should ensure compliance with employment regulations for subcontractors?
a. To avoid potential legal liabilities and penalties.
b. To ensure the subcontractor joins a union.
c. To control the subcontractor's work schedule.
d. To dictate the subcontractor's hiring practices.

Answer: a. To avoid potential legal liabilities and penalties.
Explanation: Misclassifying or not adhering to employment regulations for subcontractors can lead to legal consequences for the primary contractor.

129. In a real-world scenario, a construction company was sued for not providing equal pay for equal work among its employees. This is most likely a violation of which act?
a. The Davis-Bacon Act
b. The Fair Labor Standards Act (FLSA)
c. The Equal Pay Act
d. The Occupational Safety and Health Act (OSHA)

Answer: c. The Equal Pay Act.
Explanation: The Equal Pay Act requires that men and women be given equal pay for equal work in the same establishment.

130. A construction company that retaliates against an employee for reporting unsafe working conditions might face legal consequences under which act?
a. The Davis-Bacon Act
b. The Fair Labor Standards Act (FLSA)
c. The Occupational Safety and Health Act (OSHA)
d. The National Labor Relations Act (NLRA)

Answer: c. The Occupational Safety and Health Act (OSHA).
Explanation: OSHA prohibits employers from retaliating against employees for exercising their rights under the Act, including the right to raise health and safety concerns.

131. Case 1: ABC Construction recently hired a group of workers for a new project. Among them was Maria, a 50-year-old experienced bricklayer. However, she noticed that her younger colleagues were getting paid more for the same job.

Which law might Maria refer to if she believes she's being discriminated against based on her age?
a. The Davis-Bacon Act
b. The Age Discrimination in Employment Act (ADEA)
c. The Occupational Safety and Health Act (OSHA)
d. The National Labor Relations Act (NLRA)

Answer: b. The Age Discrimination in Employment Act (ADEA).
Explanation: The ADEA protects employees aged 40 and above from discrimination based on age.

132. Case 2: XYZ Builders has a policy of not hiring workers who have previously filed workers' compensation claims. John, who had a minor injury three years ago and filed a claim, was denied a job based on this policy.

Which action might John consider based on XYZ Builders' hiring policy?
a. Filing a complaint for violation of the Fair Labor Standards Act (FLSA).
b. Seeking damages under the Occupational Safety and Health Act (OSHA).
c. Challenging the policy as discriminatory under employment laws.
d. Requesting an audit of XYZ Builders' financial records.

Answer: c. Challenging the policy as discriminatory under employment laws.
Explanation: Discriminating against someone for having filed a workers' compensation claim can be considered retaliatory and discriminatory.

133. Case 3: DEF Construction has been employing several workers as "independent contractors," even though they work full-time hours and use company equipment.

What potential legal consequence might DEF Construction face?
a. Being required to back pay for overtime under the FLSA.
b. Penalties under the Davis-Bacon Act.
c. A lawsuit for not providing safety equipment.
d. A requirement to train all workers on union rights.

Answer: a. Being required to back pay for overtime under the FLSA.
Explanation: Misclassifying employees as independent contractors can lead to violations of wage and hour laws, including owed overtime.

134. Case 4: GHI Builders has a strict policy that all workers must speak English on the job site, even during breaks. Several Spanish-speaking workers feel this is discriminatory.

Which law might protect the rights of these workers?
a. The Davis-Bacon Act
b. The Occupational Safety and Health Act (OSHA)
c. The National Labor Relations Act (NLRA)
d. Title VII of the Civil Rights Act

Answer: d. Title VII of the Civil Rights Act.
Explanation: Title VII prohibits employment discrimination based on national origin, which can include language-based discrimination.

135. Case 5: JKL Construction recently fired an employee after he reported unsafe working conditions to OSHA.

Which act protects employees from retaliation in such scenarios?
a. The Davis-Bacon Act
b. The Whistleblower Protection Act
c. The Occupational Safety and Health Act (OSHA)
d. The Fair Labor Standards Act (FLSA)

Answer: c. The Occupational Safety and Health Act (OSHA).
Explanation: OSHA prohibits employers from retaliating against employees for reporting unsafe working conditions.

136. Case 6: MNO Construction has a policy of not providing health insurance or other benefits to part-time workers. Some of these workers consistently work 35 hours a week.

Which act might these workers refer to if they believe they should be entitled to benefits?
a. The Affordable Care Act (ACA)
b. The Davis-Bacon Act
c. The Occupational Safety and Health Act (OSHA)
d. The National Labor Relations Act (NLRA)

Answer: a. The Affordable Care Act (ACA).
Explanation: The ACA requires certain employers to offer health insurance to employees who work an average of at least 30 hours a week.

137. Case 7: PQR Builders recently faced a lawsuit from a female employee who claimed she was paid less than her male counterparts for the same job.

Which act supports her claim?
a. The Equal Pay Act
b. The Davis-Bacon Act
c. The Occupational Safety and Health Act (OSHA)
d. The National Labor Relations Act (NLRA)

Answer: a. The Equal Pay Act.
Explanation: The Equal Pay Act requires that men and women be given equal pay for equal work.

138. Case 8: STU Construction has a policy of mandatory retirement at age 65. Tom, a 67-year-old foreman, believes this policy is discriminatory.

Which act might Tom refer to in challenging this policy?
a. The Davis-Bacon Act
b. The Age Discrimination in Employment Act (ADEA)
c. The Occupational Safety and Health Act (OSHA)
d. The National Labor Relations Act (NLRA)

Answer: b. The Age Discrimination in Employment Act (ADEA).
Explanation: The ADEA prohibits employment discrimination against persons 40 years

139. Case 1: ABC Construction entered into a contract with a client to complete a building project within 12 months. Due to unforeseen circumstances, including a global pandemic, they couldn't finish the project on time. The client is now suing for breach of contract.

Which legal concept might ABC Construction invoke to defend against the lawsuit?
a. Specific performance
b. Quantum meruit
c. Force majeure
d. Estoppel

Answer: c. Force majeure.
Explanation: Force majeure refers to unforeseeable circumstances that prevent someone from fulfilling a contract. A global pandemic could be considered such a circumstance.

140. Case 2: XYZ Builders agreed to use a specific brand of tiles in a client's home renovation. However, they used a different brand, claiming it's of similar quality. The client is unhappy and wants to terminate the contract.

What type of breach has occurred?
a. Anticipatory breach
b. Material breach
c. Minor breach
d. Mutual breach

Answer: c. Minor breach.
Explanation: A minor breach (or partial breach) occurs when the essential purpose of the contract is fulfilled but some minor detail (like the brand of tiles) is not as specified.

141. Case 3: DEF Construction had a contract with a supplier for 500 bricks. The supplier delivered 450 bricks and said they would deliver the rest in a month. DEF Construction immediately terminated the contract.

Which type of breach is DEF Construction alleging?
a. Anticipatory breach
b. Material breach
c. Minor breach
d. Mutual breach

Answer: a. Anticipatory breach.
Explanation: Anticipatory breach occurs when one party indicates that they will not meet their contractual obligations.

142. Case 4: GHI Builders entered into a contract that was vague about the completion date of the project. Now, there's a dispute about the expected finish date.

Which contract principle might be in question here?
a. Offer and acceptance
b. Consideration
c. Capacity and legality
d. Clarity and specificity

Answer: d. Clarity and specificity.
Explanation: Contracts should be clear and specific to avoid ambiguities that can lead to disputes.

143. Case 5: JKL Construction signed a contract with a client. Later, they realized that the client is only 16 years old.

What might JKL Construction argue regarding the contract's validity?
a. Lack of consideration
b. Lack of capacity
c. Duress
d. Misrepresentation

Answer: b. Lack of capacity.
Explanation: Minors typically lack the legal capacity to enter into binding contracts.

144. Case 6: MNO Builders entered into a contract under the threat of physical harm. They now want to void the contract.

Which defense might MNO Builders use?
a. Mistake
b. Duress
c. Undue influence
d. Lack of consideration

Answer: b. Duress.
Explanation: Contracts entered into under duress (threats or force) can be voided.

145. Case 7: PQR Construction agreed to a contract with a client. Later, they discovered that both parties had a mistaken belief about a fundamental fact related to the project.

Which might be a valid reason to void the contract?
a. Mutual mistake
b. Unilateral mistake
c. Duress
d. Misrepresentation

Answer: a. Mutual mistake.
Explanation: A mutual mistake about a fundamental fact can be grounds to void a contract.

146. Case 8: STU Builders signed a contract that had a clause stating any disputes would be resolved through arbitration rather than court litigation.

What is this type of clause called?
a. Indemnity clause
b. Arbitration clause
c. Force majeure clause
d. Confidentiality clause

Answer: b. Arbitration clause.
Explanation: An arbitration clause specifies that disputes will be resolved through arbitration rather than court proceedings.

147. Case 9: VWX Construction was promised a bonus if they completed a project ahead of schedule. They did, but the client refused to pay the bonus.

Which legal concept might VWX Construction invoke to claim the bonus?
a. Quantum meruit
b. Specific performance
c. Estoppel
d. Rescission

Answer: b. Specific performance.
Explanation: Specific performance is a remedy where a court orders a party to perform a specific act, such as paying a promised bonus.

148. Case 10: YZA Builders entered into a contract with a client. The client later learned that YZA Builders made false statements to secure the contract.

Which legal term describes YZA Builders' actions?
a. Duress
b. Undue influence
c. Misrepresentation
d. Mutual mistake

Answer: c. Misrepresentation.
Explanation: Misrepresentation refers to making false statements to induce another party into a contract.

149. Case 1: ABC Construction was working on a high-rise building when a crane malfunctioned, causing damage to a neighboring building. Their general liability insurance was called upon to cover the damages.

Which type of coverage within the general liability insurance would address this?
a. Workers' compensation
b. Professional liability
c. Property damage liability
d. Personal injury liability

Answer: c. Property damage liability.
Explanation: Property damage liability covers damages that the policyholder's business causes to someone else's property.

150. Case 2: XYZ Builders was sued for faulty design in a residential project they completed. The homeowners claimed the design flaws led to water leakage.

Which insurance might XYZ Builders claim?
a. Commercial auto insurance
b. Professional liability insurance
c. Builders risk insurance
d. Umbrella insurance

Answer: b. Professional liability insurance.
Explanation: Professional liability insurance covers businesses against negligence claims due to harm from mistakes or failure to perform.

151. Case 3: DEF Construction's equipment was stolen from a job site. They had an insurance policy that covered the replacement cost of the equipment.

Which type of insurance did DEF Construction have?
a. Inland marine insurance
b. Commercial property insurance
c. General liability insurance
d. Employment practices liability insurance

Answer: a. Inland marine insurance.
Explanation: Inland marine insurance covers property in transit and other movable property.

152. Case 4: GHI Builders was required to purchase a bond guaranteeing they would complete a public school project. The bond would compensate the school district if GHI failed to finish the project.

What type of bond did GHI Builders obtain?
a. Payment bond
b. Bid bond
c. Performance bond
d. Maintenance bond

Answer: c. Performance bond.
Explanation: A performance bond guarantees the completion of a project and compensates the project owner if the contractor fails to complete the project as per terms.

153. Case 5: JKL Construction was bidding on a large government project. They were required to provide assurance that they would take on the project if their bid was accepted.

Which bond is relevant in this scenario?
a. Payment bond
b. Bid bond
c. Performance bond
d. Maintenance bond

Answer: b. Bid bond.
Explanation: A bid bond assures that a contractor will undertake a project if they win the bid.

154. Case 6: MNO Builders completed a project, but six months later, the client noticed some defects in the work. The client claimed against a bond that guaranteed the repair of such defects.

What type of bond was claimed?
a. Payment bond
b. Bid bond
c. Performance bond
d. Maintenance bond

Answer: d. Maintenance bond.
Explanation: Maintenance bonds guarantee against defects for a certain period after a project's completion.

155. Case 7: PQR Construction was working on a project when a fire broke out, damaging the structure. They had insurance that covered damages from such risks during construction.

Which insurance did PQR Construction have?
a. Builders risk insurance
b. Commercial property insurance
c. Professional liability insurance
d. Umbrella insurance

Answer: a. Builders risk insurance.
Explanation: Builders risk insurance covers damage to a building under construction.

156. Case 8: STU Builders was sued by a former employee for wrongful termination. The employee sought compensation for lost wages and emotional distress.

Which insurance might cover STU Builders in this situation?
a. Workers' compensation
b. Employment practices liability insurance
c. Professional liability insurance
d. Commercial auto insurance

Answer: b. Employment practices liability insurance.
Explanation: Employment practices liability insurance covers claims from employees alleging wrongful treatment.

157. Case 9: VWX Construction was involved in a major project when a hurricane hit, causing significant damage. They had a specific insurance to cover such natural disasters.

Which type of insurance did VWX Construction have?
a. Commercial property insurance
b. Builders risk insurance with a natural disasters clause
c. Inland marine insurance
d. Umbrella insurance

Answer: b. Builders risk insurance with a natural disasters clause.
Explanation: Builders risk insurance can have specific clauses to cover natural disasters like hurricanes.

158. Case 10: YZA Builders was transporting construction equipment to a site when the truck met with an accident, damaging the equipment.

Which insurance might cover the damages?
a. Inland marine insurance
b. Commercial auto insurance
c. General liability insurance
d. Professional liability insurance

Answer: a. Inland marine insurance.
Explanation: Inland marine insurance covers equipment and other property while in transit.

159. Which insurance provides coverage for injuries or property damage a construction company might cause to third parties?
a. Professional liability insurance
b. Workers' compensation insurance
c. General liability insurance
d. Builders risk insurance

Answer: c. General liability insurance.
Explanation: General liability insurance covers claims of bodily injury, property damage, and personal injury to third parties caused by the contractor's operations or products.

160. A construction company is sued for alleged negligence in providing design services. Which insurance would most likely cover this claim?
a. Commercial auto insurance
b. General liability insurance
c. Professional liability insurance
d. Workers' compensation insurance

Answer: c. Professional liability insurance.
Explanation: Professional liability insurance, often referred to as errors and omissions (E&O) insurance, covers negligence claims related to professional services provided.

161. Which insurance covers a construction company's employees if they get injured on the job?
a. General liability insurance
b. Professional liability insurance
c. Workers' compensation insurance
d. Commercial auto insurance

Answer: c. Workers' compensation insurance.
Explanation: Workers' compensation insurance provides wage replacement and medical benefits to employees injured in the course of employment.

162. A construction company's equipment is damaged in a flood while stored at the job site. Which insurance might cover this loss?
a. Commercial auto insurance
b. General liability insurance
c. Builders risk insurance
d. Professional liability insurance

Answer: c. Builders risk insurance.
Explanation: Builders risk insurance covers damage to a building under construction and may include coverage for equipment and materials on site.

163. Which insurance would cover a construction company if one of its trucks is involved in an accident?
a. Commercial auto insurance
b. General liability insurance
c. Professional liability insurance
d. Workers' compensation insurance

Answer: a. Commercial auto insurance.
Explanation: Commercial auto insurance covers vehicles owned and used by a business, including trucks used in construction.

164. Why is general liability insurance crucial for construction contractors?
a. It covers employee injuries on the job.
b. It covers design flaws and professional errors.
c. It protects against claims of property damage or bodily injury caused by the contractor's operations.
d. It covers damages to the contractor's equipment.

Answer: c. It protects against claims of property damage or bodily injury caused by the contractor's operations.
Explanation: General liability insurance is essential for contractors as it provides coverage against third-party claims arising from the contractor's operations, such as property damage or bodily injury.

165. Which insurance would a contractor purchase to ensure coverage against claims of discrimination or wrongful termination?
a. Employment practices liability insurance
b. General liability insurance
c. Professional liability insurance
d. Workers' compensation insurance

Answer: a. Employment practices liability insurance. Explanation: Employment practices liability insurance covers claims from employees alleging wrongful treatment, such as discrimination or wrongful termination.

166. In the event of an employee's long-term injury, which insurance provides rehabilitation and covers a portion of the injured employee's lost wages?
a. General liability insurance
b. Professional liability insurance
c. Workers' compensation insurance
d. Commercial auto insurance

Answer: c. Workers' compensation insurance. Explanation: Workers' compensation insurance provides medical benefits, rehabilitation, and a portion of lost wages to employees injured on the job.

167. A subcontractor hired by a general contractor causes damage to a client's property. Under which insurance might the general contractor seek coverage, even if the fault was the subcontractor's?
a. Professional liability insurance
b. Workers' compensation insurance
c. General liability insurance
d. Commercial auto insurance

Answer: c. General liability insurance. Explanation: The general contractor's general liability insurance might cover damages caused by subcontractors, depending on the policy's terms.

168. A construction company wants to ensure coverage beyond the limits of their standard policies. Which type of insurance provides additional liability coverage above the limits of an insured's primary policies?
a. Umbrella insurance
b. General liability insurance
c. Professional liability insurance
d. Workers' compensation insurance

Answer: a. Umbrella insurance.
Explanation: Umbrella insurance provides additional liability coverage beyond the limits of the primary policies, offering an extra layer of protection.

169. A construction company is bidding on a project and wants to ensure the owner that they can secure the necessary performance and payment bonds if awarded the contract. Which type of bond should they obtain?
a. Maintenance bond
b. Payment bond
c. Bid bond
d. Performance bond

Answer: c. Bid bond.
Explanation: A bid bond assures the project owner that the bidder can obtain the required performance and payment bonds if awarded the contract.

170. Which bond guarantees that a contractor will pay for labor and materials used in a project?
a. Bid bond
b. Performance bond
c. Payment bond
d. Maintenance bond

Answer: c. Payment bond. Explanation: A payment bond ensures that the contractor will pay for services, labor, and materials used in the project.

171. A project owner wants to ensure that the contractor will complete the project according to the terms of the contract. Which bond provides this assurance?
a. Bid bond
b. Performance bond
c. Payment bond
d. Supply bond

Answer: b. Performance bond.
Explanation: A performance bond guarantees that the contractor will perform the work according to the contract's terms and conditions.

172. In the event a contractor defaults on a project, which bond can compensate the project owner for financial losses?
a. Bid bond
b. Performance bond
c. Payment bond
d. Maintenance bond

Answer: b. Performance bond.
Explanation: A performance bond provides financial compensation to the project owner if the contractor defaults or fails to perform according to the contract.

173. Which bond ensures that any defects found after project completion will be repaired by the contractor?
a. Bid bond
b. Performance bond
c. Payment bond
d. Maintenance bond

Answer: d. Maintenance bond.
Explanation: A maintenance bond guarantees that the contractor will correct any defects or faulty work discovered after the project's completion.

174. When a contractor is seeking to obtain a bond, which party typically underwrites and issues the bond?
a. The project owner
b. The contractor
c. The subcontractor
d. A surety company

Answer: d. A surety company.
Explanation: Bonds are typically underwritten and issued by a surety company on behalf of the contractor.

175. Which of the following is NOT a typical consequence for contractors who fail to meet insurance and bonding requirements?
a. Fines and penalties
b. Loss of licensing
c. Increased project bids
d. Legal action

Answer: c. Increased project bids. Explanation: While failing to meet insurance and bonding requirements can lead to fines, loss of licensing, and legal action, it doesn't directly result in increased project bids.

176. A contractor was unable to complete a project due to financial difficulties, and the project owner had to hire another contractor to finish the job. Which bond would the project owner call upon to cover the additional costs?
a. Bid bond
b. Performance bond
c. Payment bond
d. Maintenance bond

Answer: b. Performance bond. Explanation: A performance bond provides financial protection to the project owner if the contractor defaults or fails to complete the project.

177. In the construction industry, why is it essential for contractors to maintain both insurance and bonds?
a. To increase their profit margins
b. To ensure they can bid on any project
c. To provide financial protection and guarantee their contractual obligations
d. To reduce the need for subcontractors

Answer: c. To provide financial protection and guarantee their contractual obligations. Explanation: Insurance provides financial protection against potential losses, while bonds guarantee the contractor's contractual obligations to the project owner.

178. A contractor failed to pay a subcontractor and several material suppliers on a project. Despite the project owner having paid the contractor in full, these parties made claims for their unpaid dues. Which bond would cover these claims?
a. Bid bond
b. Performance bond
c. Payment bond
d. Maintenance bond

Answer: c. Payment bond.
Explanation: A payment bond ensures that the contractor will pay for services, labor, and materials, protecting parties like subcontractors and suppliers from non-payment.

179. A contractor has secured a performance bond for a project. How might this influence the contractor-client relationship?
a. It may reduce the client's trust in the contractor.
b. It provides the client with assurance that the project will be completed as per the contract.
c. It increases the project's overall cost for the client.
d. It limits the contractor's ability to make changes to the project.

Answer: b. It provides the client with assurance that the project will be completed as per the contract.
Explanation: A performance bond acts as a guarantee that the contractor will fulfill the project's terms, providing the client with added confidence.

180. Which of the following is a primary reason for a construction company to review its insurance and bonding coverage annually?
a. To ensure compliance with changing regulations.
b. To reduce the premiums they pay.
c. To adjust for inflation.
d. To account for changes in company size and project scope.

Answer: d. To account for changes in company size and project scope.
Explanation: As a company grows or takes on different projects, its risk profile changes, necessitating a review of insurance and bonding coverage.

181. A subcontractor working on a project accidentally damages a client's property. Under which insurance would this damage typically be covered?
a. Workers' Compensation Insurance
b. Professional Liability Insurance
c. General Liability Insurance
d. Performance Bond

Answer: c. General Liability Insurance.
Explanation: General Liability Insurance covers property damage and bodily injury caused during the course of work.

182. Which of the following is NOT a typical method for a construction company to ensure they have adequate insurance and bonding coverage?
a. Regularly reviewing and updating company risk assessments.
b. Consulting with an insurance broker specializing in the construction industry.
c. Relying solely on historical data without considering future projects.
d. Engaging in periodic audits of insurance and bond policies.

Answer: c. Relying solely on historical data without considering future projects.
Explanation: While historical data is valuable, relying solely on it without considering future projects and growth can lead to inadequate coverage.

183. A client is concerned about potential injuries to workers on their property during a construction project. Which insurance should the contractor have to address this concern?
a. Professional Liability Insurance
b. General Liability Insurance
c. Workers' Compensation Insurance
d. Payment Bond

Answer: c. Workers' Compensation Insurance.
Explanation: Workers' Compensation Insurance covers injuries to employees while on the job, ensuring they receive medical care and compensation for lost wages.

184. How can a construction company ensure that subcontractors and independent contractors have adequate insurance coverage?
a. By requesting and reviewing certificates of insurance.
b. By relying on verbal assurances.
c. By checking online reviews of the subcontractor.
d. By assuming that all subcontractors have the necessary coverage.

Answer: a. By requesting and reviewing certificates of insurance.
Explanation: Certificates of insurance provide proof of coverage and allow the construction company to verify that subcontractors and independent contractors have the necessary insurance.

185. In the event of a dispute over work quality, which bond can a client call upon to ensure the contractor rectifies the issue or compensates for the inadequate work?
a. Bid Bond
b. Payment Bond
c. Performance Bond
d. Maintenance Bond

Answer: c. Performance Bond.
Explanation: A performance bond ensures that the contractor will complete the project according to the contract's terms, including work quality.

186. Which of the following is a potential consequence for contractors who do not maintain adequate insurance and bonding coverage?
a. Reduced project timelines
b. Inability to secure large contracts or work with certain clients
c. Increased profit margins
d. Faster approval of permits

Answer: b. Inability to secure large contracts or work with certain clients.
Explanation: Many clients and large projects require contractors to have specific insurance and bonding coverage as a prerequisite.

187. An independent contractor is injured while working on a construction site but does not have personal insurance. Who is typically responsible for the medical expenses?
a. The project owner
b. The primary contractor
c. The injured party
d. The supplier of the materials

Answer: c. The injured party.
Explanation: Independent contractors are typically responsible for their insurance. If they don't have coverage, they may be personally liable for medical expenses.

188. A construction company is expanding its operations and taking on larger projects. What step should they prioritize regarding insurance and bonds?
a. Reducing coverage to save on premiums.
b. Consulting with an industry-specific insurance broker to reassess their needs.
c. Ignoring bonds as they are not mandatory.
d. Relying on existing policies without review.

Answer: b. Consulting with an industry-specific insurance broker to reassess their needs.
Explanation: As a company grows and takes on larger projects, it's crucial to reassess insurance and bonding needs to ensure adequate coverage.

189. Case Study 1: ABC Construction has recently expanded its operations and hired several new employees. The company's management is considering implementing a new project management software to streamline operations.

Which of the following considerations is LEAST relevant when selecting a project management software?
a. The software's compatibility with the company's current IT infrastructure.
b. The popularity of the software among other construction companies.
c. The software's ability to integrate with other tools the company uses.
d. The training and support provided by the software vendor.

Answer: b. The popularity of the software among other construction companies.
Explanation: While popularity might indicate a well-liked product, it's more important to ensure the software meets the specific needs of ABC Construction.

190. Case Study 2: XYZ Builders has noticed a decline in productivity and an increase in project delays. After an internal audit, it was found that communication breakdowns between departments were a significant factor.

Which strategy would be MOST effective in improving inter-departmental communication?
a. Implementing a company-wide communication platform.
b. Increasing the number of company-wide meetings.
c. Hiring more employees to manage the workload.
d. Investing in individual training for each department.

Answer: a. Implementing a company-wide communication platform.
Explanation: A unified communication platform can streamline communication, ensuring all departments are aligned and informed.

191. Case Study 3: DEF Construction is facing financial difficulties due to unpaid client invoices. They are considering options to improve their cash flow.

Which of the following would be the MOST effective approach to address this issue?
a. Taking on more projects to increase revenue.
b. Offering discounts for early payment of invoices.
c. Reducing the number of employees to cut costs.
d. Investing in expensive marketing campaigns.

Answer: b. Offering discounts for early payment of invoices.
Explanation: Encouraging clients to pay their invoices early by offering discounts can improve cash flow in the short term.

192. Case Study 4: GHI Contractors has received feedback that their clients find their contract terms confusing, leading to misunderstandings and disputes.

What should GHI Contractors prioritize to address this issue?
a. Reducing the length of their contracts.
b. Using more technical jargon to appear professional.
c. Simplifying and clarifying contract language.
d. Increasing the prices to compensate for potential disputes.

Answer: c. Simplifying and clarifying contract language.
Explanation: Clear and straightforward contract terms reduce misunderstandings and foster better client relationships.

193. Case Study 5: JKL Renovations is considering expanding into a new region. They are weighing the potential benefits against the risks.

Which of the following is NOT a primary consideration when expanding into a new region?
a. The popularity of the company's brand in the new region.
b. The regulatory and licensing requirements of the new region.
c. The cultural and architectural preferences of the new region.
d. The favorite color of the company's CEO.

Answer: d. The favorite color of the company's CEO.
Explanation: Personal preferences of executives, like favorite colors, are irrelevant when making strategic business decisions.

194. Case Study 6: MNO Builders has a high employee turnover rate. Exit interviews indicate dissatisfaction with management as a common reason for leaving.

Which strategy would likely be MOST effective in retaining employees?
a. Offering higher salaries than competitors.
b. Implementing regular team-building exercises.
c. Providing management training to improve leadership skills.
d. Hiring employees on a short-term contract basis.

Answer: c. Providing management training to improve leadership skills.
Explanation: Addressing the root cause, which is dissatisfaction with management, through leadership training can improve employee retention.

195. Case Study 7: PQR Construction is considering adopting sustainable building practices. They are evaluating the potential benefits and challenges.

Which of the following is a potential benefit of adopting sustainable building practices?
a. Reduced initial construction costs.
b. Enhanced company reputation and potential for new business.
c. Elimination of the need for permits and regulations.
d. Immediate financial returns on sustainable investments.

Answer: b. Enhanced company reputation and potential for new business.
Explanation: Sustainable practices can enhance a company's reputation, making them more attractive to eco-conscious clients.

196. Case Study 8: STU Developers is a family-owned business. The current owner is retiring, and there's uncertainty about the company's leadership transition.

What should STU Developers prioritize to ensure a smooth leadership transition?
a. Selling the company to the highest bidder.
b. Appointing the youngest family member as the new leader.
c. Developing a clear succession plan with input from key stakeholders.
d. Changing the company's business model entirely.

Answer: c. Developing a clear succession plan with input from key stakeholders.
Explanation: A well-thought-out succession plan ensures continuity and stability during leadership transitions.

197. Case Study 9: VWX Construction has been receiving complaints about the quality of their work. They are considering ways to address this issue.

Which of the following would be the MOST effective way to improve work quality?
a. Investing in advertising to counteract negative feedback.
b. Implementing rigorous quality control checks and training.
c. Reducing prices to make up for the quality issues.
d. Ignoring the feedback and hoping it goes away.

Answer: b. Implementing rigorous quality control checks and training.
Explanation: Addressing the root cause of the issue, which is work quality, through checks and training will lead to long-term improvements.

198. Case Study 10: YZA Contractors is considering diversifying their services to include interior design. They are evaluating the potential risks and benefits.

Which of the following is a potential risk of diversifying their services?
a. Increased market visibility and brand recognition.
b. Spreading company resources too thin.
c. Reducing the company's dependency on a single revenue stream.
d. Attracting a broader client base.

Answer: b. Spreading company resources too thin.
Explanation: While diversification can offer many benefits, it can also strain a company's resources if not managed properly.

199. Which business structure is characterized by a single owner who is responsible for all debts and obligations of the business?
a. Corporation
b. Limited Liability Company (LLC)
c. Partnership
d. Sole Proprietorship

Answer: d. Sole Proprietorship
Explanation: A sole proprietorship is owned by a single individual who is responsible for all aspects of the business, including its debts and obligations.

200. In which business structure do owners have limited personal liability for the debts and actions of the company?
a. Sole Proprietorship
b. General Partnership
c. Corporation
d. Joint Venture

Answer: c. Corporation
Explanation: In a corporation, the owners (shareholders) have limited personal liability for the company's debts and actions.

201. Which of the following is a primary advantage of forming a Limited Liability Company (LLC) in the construction industry?
a. No need for annual meetings or record-keeping.
b. Owners can avoid double taxation.
c. Unlimited personal liability for business debts.
d. The company can issue stock to raise capital.

Answer: b. Owners can avoid double taxation.
Explanation: An LLC offers pass-through taxation, meaning the company itself is not taxed. Instead, income is reported on the owners' individual tax returns.

202. A construction company formed by two individuals who share profits, losses, and responsibility is known as:
a. Corporation
b. Sole Proprietorship
c. Joint Venture
d. Partnership

Answer: d. Partnership
Explanation: A partnership involves two or more individuals who share in the profits, losses, and responsibilities of the business.

203. Which business structure allows for easy transfer of ownership and the ability to raise capital by issuing stock?
a. Sole Proprietorship
b. Partnership
c. Limited Liability Company (LLC)
d. Corporation

Answer: d. Corporation
Explanation: Corporations can issue stock, allowing them to raise capital and easily transfer ownership.

204. When creating a business plan for a construction company, which section outlines the company's financial projections and funding requirements?
a. Market Analysis
b. Organization and Management
c. Service or Product Line
d. Financial Projections

Answer: d. Financial Projections
Explanation: The Financial Projections section of a business plan details the company's expected financial performance and any funding needs.

205. In a business plan for a construction company, which section provides information about the company's legal structure and ownership?
a. Executive Summary
b. Organization and Management
c. Market Analysis
d. Service or Product Line

Answer: b. Organization and Management
Explanation: The Organization and Management section details the company's legal structure, ownership, and key management roles.

206. Which section of a construction company's business plan would detail the specific services offered, such as residential construction, commercial construction, or renovation?
a. Executive Summary
b. Market Analysis
c. Service or Product Line
d. Financial Projections

Answer: c. Service or Product Line
Explanation: The Service or Product Line section describes the specific services or products the company offers.

207. In a business plan, which section provides an overview of the construction industry's current market trends, target market, and competition?
a. Executive Summary
b. Organization and Management
c. Market Analysis
d. Service or Product Line

Answer: c. Market Analysis
Explanation: The Market Analysis section provides insights into the industry's trends, the company's target market, and its competition.

208. For a construction company looking to secure funding, which section of the business plan would be MOST crucial to potential investors or lenders?
a. Executive Summary
b. Organization and Management
c. Service or Product Line
d. Financial Projections

Answer: d. Financial Projections
Explanation: Financial Projections provide potential investors or lenders with an understanding of the company's expected financial performance and its ability to repay loans or provide returns on investments.

209. Which accounting principle requires revenue to be recorded when earned, regardless of when payment is received?
a. Cash Basis Accounting
b. Accrual Basis Accounting
c. Historical Cost Principle
d. Economic Entity Assumption

Answer: b. Accrual Basis Accounting. Explanation: Accrual Basis Accounting recognizes revenue when it's earned and expenses when they're incurred, regardless of the timing of cash flows.

210. In construction accounting, which method matches revenues to the specific costs of a particular job, allowing for a more accurate picture of job profitability?
a. Job Costing Method
b. Period Costing Method
c. Process Costing Method
d. Standard Costing Method

Answer: a. Job Costing Method
Explanation: Job Costing Method allocates specific costs to individual jobs or projects, providing a detailed view of profitability for each job.

211. Which accounting principle suggests that a construction company should record its property or equipment at the price it paid, rather than its current market value?
a. Going Concern Assumption
b. Historical Cost Principle
c. Revenue Recognition Principle
d. Monetary Unit Assumption

Answer: b. Historical Cost Principle
Explanation: The Historical Cost Principle dictates that assets should be recorded based on their original cost, not current market value.

212. For a construction company, which financial statement provides a snapshot of its financial position at a specific point in time, detailing assets, liabilities, and equity?
a. Income Statement
b. Cash Flow Statement
c. Statement of Retained Earnings
d. Balance Sheet

Answer: d. Balance Sheet
Explanation: The Balance Sheet provides a snapshot of a company's financial position, detailing its assets, liabilities, and equity at a specific point in time.

213. Which of the following is NOT typically considered a direct cost in construction job costing?
a. Labor costs for workers on the job
b. Materials used for the job
c. Equipment rental for the job
d. Office administrative salaries

Answer: d. Office administrative salaries
Explanation: Office administrative salaries are typically considered indirect costs, as they don't directly relate to a specific job.

214. A construction company that recognizes revenue only when cash is received and records expenses only when cash is paid is using which accounting method?
a. Accrual Basis Accounting
b. Cash Basis Accounting
c. Job Costing Method
d. Period Costing Method

Answer: b. Cash Basis Accounting
Explanation: Cash Basis Accounting recognizes revenue and expenses only when cash changes hands.

215. XYZ Construction, a successful construction company, operates as a corporation but is taxed as a partnership, avoiding double taxation. What business structure is XYZ Construction likely using?
a. Sole Proprietorship
b. General Partnership
c. S Corporation
d. C Corporation

Answer: c. S Corporation
Explanation: An S Corporation combines the limited liability of a corporation with the pass-through taxation of a partnership.

216. In the context of a construction company's financial reporting, which statement provides detailed information about cash inflows and outflows over a specific period?
a. Income Statement
b. Balance Sheet
c. Cash Flow Statement
d. Statement of Retained Earnings

Answer: c. Cash Flow Statement. Explanation: The Cash Flow Statement provides a detailed view of a company's cash inflows and outflows over a specific period.

217. Which accounting principle assumes that a construction company will continue its operations indefinitely and is not likely to liquidate in the foreseeable future?
a. Historical Cost Principle
b. Going Concern Assumption
c. Revenue Recognition Principle
d. Monetary Unit Assumption

Answer: b. Going Concern Assumption
Explanation: The Going Concern Assumption assumes that a business will continue to operate indefinitely unless there's evidence to the contrary.

218. For a construction company, which of the following would be considered an indirect cost when estimating the cost of a specific job?
a. Cost of bricks for a wall
b. Salary of the project manager
c. Rent for the company's main office
d. Wages of the laborers working on the job

Answer: c. Rent for the company's main office
Explanation: Rent for the company's main office is an indirect cost, as it doesn't directly relate to a specific job but is spread across multiple projects.

219. Which of the following is a primary reason for a construction contractor to understand basic accounting principles?
a. To ensure compliance with building codes
b. To effectively manage and track financial performance
c. To design construction projects
d. To communicate with subcontractors

Answer: b. To effectively manage and track financial performance
Explanation: Understanding basic accounting principles allows contractors to monitor the financial health of their business, make informed decisions, and ensure profitability.

220. In strategic planning for a construction company, what is the primary purpose of setting clear, measurable goals?
a. To ensure all employees are busy
b. To provide a direction and benchmark for success
c. To impress stakeholders and clients
d. To fulfill regulatory requirements

Answer: b. To provide a direction and benchmark for success
Explanation: Setting clear, measurable goals provides a roadmap for the company and allows for the assessment of performance against these goals.

221. Which component of a construction company's accounting system categorizes and organizes financial transactions?
a. Job costing
b. Financial reporting
c. Chart of accounts
d. Performance metrics

Answer: c. Chart of accounts
Explanation: The chart of accounts is a listing of all accounts in the accounting system, organizing transactions into categories like assets, liabilities, revenue, and expenses.

222. For a construction company, which financial statement would best indicate the profitability of the company over a specific period?
a. Balance Sheet
b. Cash Flow Statement
c. Income Statement
d. Statement of Retained Earnings

Answer: c. Income Statement
Explanation: The Income Statement, also known as the Profit and Loss Statement, provides a summary of revenues, costs, and expenses, indicating the profitability over a specific period.

223. In the context of financial management for construction, what does job costing primarily help with?
a. Determining the company's overall profitability
b. Assessing the financial health of subcontractors
c. Allocating costs to specific projects to determine project profitability
d. Setting long-term strategic goals

Answer: c. Allocating costs to specific projects to determine project profitability
Explanation: Job costing allocates specific costs to individual projects, allowing for a detailed view of each project's profitability.

224. Which of the following is NOT typically a component of a construction company's strategic plan?
a. Detailed construction blueprints
b. Vision and mission statements
c. SWOT analysis (Strengths, Weaknesses, Opportunities, Threats)
d. Short-term and long-term goals

Answer: a. Detailed construction blueprints
Explanation: While blueprints are essential for construction projects, they are not typically part of a company's strategic plan, which focuses on broader business goals and strategies.

225. When setting up the accounting system for a new construction company, which step should be prioritized first?
a. Hiring an external auditor
b. Setting up a chart of accounts tailored to the construction industry
c. Preparing the first annual report
d. Calculating the first year's profit

Answer: b. Setting up a chart of accounts tailored to the construction industry
Explanation: The chart of accounts is foundational in an accounting system, organizing and categorizing financial transactions.

226. Which of the following best describes the role of financial reporting in a construction company?
a. To provide a detailed design plan for construction projects
b. To communicate the company's financial performance to stakeholders
c. To ensure all workers are trained in safety protocols
d. To purchase construction materials

Answer: b. To communicate the company's financial performance to stakeholders
Explanation: Financial reporting provides a summary of the company's financial activities and health, communicating this information to stakeholders like owners, investors, and lenders.

227. In the construction industry, why is it crucial to differentiate between direct and indirect costs in job costing?
a. To ensure compliance with building codes
b. To accurately assess the profitability of specific projects
c. To determine the company's marketing budget
d. To communicate with clients about design preferences

Answer: b. To accurately assess the profitability of specific projects
Explanation: Differentiating between direct costs (specific to a project) and indirect costs (spread across multiple projects) ensures accurate allocation of costs and a clear understanding of project profitability.

228. For a construction company, which of the following would be considered an indirect cost?
a. Cost of bricks for a specific project
b. Wages of laborers working on a specific project
c. Rent for the company's main office
d. Cost of cement for a specific project

Answer: c. Rent for the company's main office
Explanation: Rent for the company's main office is an indirect cost, as it doesn't directly relate to a specific project but is spread across multiple projects.

229. Which of the following best describes the primary purpose of a cash flow statement in the construction industry?
a. To detail the company's assets and liabilities
b. To outline the company's revenue and expenses over a period
c. To show the movement of cash in and out of the business
d. To list all the company's stakeholders

Answer: c. To show the movement of cash in and out of the business
Explanation: The cash flow statement provides a detailed view of how cash is moving within the business, highlighting the inflows and outflows over a specific period.

230. In budgeting for a construction project, which factor is crucial to consider for ensuring profitability?
a. The company's marketing strategies
b. The CEO's salary
c. Direct and indirect costs associated with the project
d. The number of competitors in the market

Answer: c. Direct and indirect costs associated with the project
Explanation: Accurately accounting for both direct (specific to the project) and indirect costs (spread across projects) is essential to determine the project's profitability.

231. Which accounting principle dictates that expenses should be matched with the revenues they helped generate?
a. Going Concern Principle
b. Historical Cost Principle
c. Matching Principle
d. Revenue Recognition Principle

Answer: c. Matching Principle
Explanation: The Matching Principle ensures that expenses are reported in the same period as the revenues they helped to earn, providing a clearer picture of profitability.

232. For a construction company, which financial metric would be most useful in assessing the company's ability to meet short-term obligations?
a. Return on Equity
b. Gross Profit Margin
c. Current Ratio
d. Debt to Equity Ratio

Answer: c. Current Ratio
Explanation: The Current Ratio, calculated as current assets divided by current liabilities, indicates a company's ability to pay off its short-term debts.

233. When analyzing the financial health of a construction company, which statement provides insights into the company's profitability over a specific period?
a. Balance Sheet
b. Income Statement
c. Cash Flow Statement
d. Statement of Owner's Equity

Answer: b. Income Statement
Explanation: The Income Statement, or Profit and Loss Statement, provides a summary of the company's revenues, costs, and expenses over a specific period, indicating profitability.

234. In the context of cash flow management, why is it essential for construction companies to monitor their accounts receivable closely?
a. To ensure compliance with building codes
b. To maintain a steady cash inflow and manage liquidity
c. To assess the company's long-term investments
d. To calculate the company's tax obligations

Answer: b. To maintain a steady cash inflow and manage liquidity
Explanation: Monitoring accounts receivable ensures that clients are paying on time, which is crucial for maintaining liquidity and managing cash flow.

235. Which of the following is a primary reason for a construction company to set a contingency budget?
a. To account for potential unexpected costs during a project
b. To allocate funds for the company's annual retreat
c. To invest in stock market
d. To pay dividends to shareholders

Answer: a. To account for potential unexpected costs during a project
Explanation: A contingency budget is set aside to cover unforeseen expenses that might arise during a construction project, ensuring the project remains on track financially.

236. Which accounting principle states that businesses should record transactions at their original cost?
a. Revenue Recognition Principle
b. Going Concern Principle
c. Historical Cost Principle
d. Matching Principle

Answer: c. Historical Cost Principle
Explanation: The Historical Cost Principle dictates that businesses should record assets at their original cost, ensuring consistency and comparability in financial statements.

237. For a construction company, which of the following would be considered a fixed cost?
a. Cost of construction materials for a specific project
b. Wages of laborers working on a specific project
c. Monthly rent for the company's office space
d. Fuel for construction vehicles for a specific project

Answer: c. Monthly rent for the company's office space
Explanation: Monthly rent remains consistent regardless of the number of projects undertaken, making it a fixed cost.

238. In financial analysis, which metric would be most useful for a construction company to understand its profitability relative to its total sales?
a. Current Ratio
b. Debt to Equity Ratio
c. Gross Profit Margin
d. Quick Ratio

Answer: c. Gross Profit Margin
Explanation: Gross Profit Margin, calculated as (Gross Profit / Sales) x 100, indicates the percentage of total sales that exceeds the cost of goods sold, providing insights into profitability.

239. A construction company is considering taking on a large project. Which financial document would be most useful in determining if the company has enough liquidity to cover the project's initial costs?
a. Income Statement
b. Balance Sheet
c. Cash Flow Statement
d. Statement of Owner's Equity

Answer: c. Cash Flow Statement
Explanation: The Cash Flow Statement provides a detailed view of the company's liquidity by showing the inflows and outflows of cash over a specific period.

240. Which financial ratio would be most relevant for a construction company to assess its long-term solvency?
a. Gross Profit Margin
b. Current Ratio
c. Debt to Equity Ratio
d. Quick Ratio

Answer: c. Debt to Equity Ratio
Explanation: The Debt to Equity Ratio indicates the proportion of equity and debt the company is using to finance its assets, providing insights into its long-term solvency.

241. In the context of financial management, what is the primary purpose of job costing for a construction company?
a. To determine the company's overall profitability
b. To assess the financial performance of individual projects
c. To calculate the company's tax obligations
d. To set the company's strategic goals

Answer: b. To assess the financial performance of individual projects
Explanation: Job costing allows a construction company to track costs and revenues associated with specific projects, ensuring each project is financially viable.

242. When preparing a budget for a construction project, which of the following would be considered a variable cost?
a. Monthly office rent
b. CEO's annual salary
c. Cost of construction materials
d. Insurance premium for company vehicles

Answer: c. Cost of construction materials
Explanation: The cost of construction materials can vary based on the specific requirements of each project, making it a variable cost.

243. Which financial metric would be most useful for a construction company to understand its efficiency in using assets to generate profit?
a. Gross Profit Margin
b. Return on Assets (ROA)
c. Current Ratio
d. Debt to Equity Ratio

Answer: b. Return on Assets (ROA)
Explanation: ROA indicates how effectively a company's assets are being used to generate profit, providing insights into operational efficiency.

244. In financial management, which of the following best describes the concept of "working capital" for a construction company?
a. Total assets minus total liabilities
b. Current assets minus current liabilities
c. Gross profit minus operating expenses
d. Total revenue minus cost of goods sold

Answer: b. Current assets minus current liabilities
Explanation: Working capital represents the short-term liquidity of a company, indicating its ability to cover its short-term obligations.

245. For a construction company, which of the following would be a key consideration when determining its capital structure?
a. Marketing strategies
b. The mix of debt and equity financing
c. The company's mission and vision statements
d. The number of employees

Answer: b. The mix of debt and equity financing
Explanation: The capital structure refers to how a company finances its operations and growth, typically through a mix of debt and equity.

246. Which of the following best describes the purpose of a pro forma financial statement in the construction industry?
a. To report the company's historical financial performance
b. To project future financial performance based on certain assumptions
c. To detail the company's tax obligations
d. To list all the company's stakeholders

Answer: b. To project future financial performance based on certain assumptions
Explanation: Pro forma financial statements are used to forecast a company's financial performance, helping in planning and decision-making.

247. A construction company is considering investing in new machinery. Which financial analysis technique would be most appropriate to determine the potential return on this investment?
a. Current Ratio analysis
b. Gross Profit Margin calculation
c. Net Present Value (NPV) analysis
d. Debt to Equity Ratio analysis

Answer: c. Net Present Value (NPV) analysis
Explanation: NPV analysis helps in determining the potential profitability of an investment by comparing the present value of expected future cash flows to the initial investment.

248. In the context of financial management for a construction company, which of the following best describes the concept of "overhead"?
a. Direct costs associated with specific projects
b. Indirect costs not tied to a specific project but necessary for business operations
c. Revenue generated from primary business activities
d. Profit after all expenses have been deducted

Answer: b. Indirect costs not tied to a specific project but necessary for business operations
Explanation: Overhead refers to the ongoing business expenses not directly attributed to creating a product or service but essential for the overall business operations.

249. Which financial statement provides a snapshot of a construction company's assets, liabilities, and equity at a specific point in time?
a. Income Statement
b. Cash Flow Statement
c. Balance Sheet
d. Statement of Retained Earnings

Answer: c. Balance Sheet
Explanation: The Balance Sheet provides a detailed view of a company's financial position at a specific moment, detailing its assets, liabilities, and equity.

250. A construction company wants to understand its profitability over the past fiscal year. Which financial statement should it primarily refer to?
a. Balance Sheet
b. Cash Flow Statement
c. Income Statement
d. Statement of Changes in Equity

Answer: c. Income Statement
Explanation: The Income Statement provides a detailed view of a company's revenues, expenses, and profits over a specific period.

251. Which financial statement would be most useful for a construction company to assess its ability to generate cash from its operations?
a. Income Statement
b. Statement of Changes in Equity
c. Balance Sheet
d. Cash Flow Statement

Answer: d. Cash Flow Statement. Explanation: The Cash Flow Statement tracks the inflows and outflows of cash from various activities, including operations, investments, and financing.

252. A construction company's owner wants to understand how much of the profit was retained in the business versus distributed as dividends. Which statement would provide this information?
a. Income Statement
b. Cash Flow Statement
c. Balance Sheet
d. Statement of Retained Earnings

Answer: d. Statement of Retained Earnings
Explanation: The Statement of Retained Earnings provides details on changes in the company's retained earnings over a period, including profits earned and dividends distributed.

253. Which section of the Cash Flow Statement would a construction company look at to see the cash spent on purchasing new equipment?
a. Operating Activities
b. Financing Activities
c. Investing Activities
d. Non-operating Activities

Answer: c. Investing Activities
Explanation: The Investing Activities section of the Cash Flow Statement tracks cash flows related to investments, such as the purchase or sale of long-term assets like equipment.

254. When analyzing the Income Statement, which metric would be most relevant for a construction company to understand its core profitability?
a. Gross Profit Margin
b. Current Ratio
c. Return on Equity
d. Debt to Asset Ratio

Answer: a. Gross Profit Margin
Explanation: Gross Profit Margin indicates the percentage of revenue that exceeds the cost of goods sold, reflecting the core profitability from primary operations.

255. Which financial statement would be most relevant for a construction company's stakeholders to assess its short-term liquidity?
a. Income Statement
b. Balance Sheet
c. Cash Flow Statement
d. Statement of Retained Earnings

Answer: b. Balance Sheet
Explanation: The Balance Sheet provides insights into a company's current assets and liabilities, helping stakeholders assess short-term liquidity.

256. A construction company is considering taking on debt to finance a new project. Which financial statement section would detail its current liabilities?
a. Income Statement's "Expenses" section
b. Cash Flow Statement's "Financing Activities" section
c. Balance Sheet's "Liabilities" section
d. Statement of Retained Earnings' "Dividends" section

Answer: c. Balance Sheet's "Liabilities" section
Explanation: The Balance Sheet's "Liabilities" section provides a detailed view of the company's current and long-term obligations.

257. For a construction company, which financial statement would best indicate the revenue generated from its primary business activities over the past quarter?
a. Balance Sheet
b. Cash Flow Statement
c. Income Statement
d. Statement of Changes in Equity

Answer: c. Income Statement
Explanation: The Income Statement provides a detailed view of a company's revenues and expenses over a specific period, indicating the revenue from primary business activities.

258. A construction company's stakeholder wants to understand the net change in cash over the past year. Which financial statement would provide this information?
a. Income Statement
b. Cash Flow Statement
c. Balance Sheet
d. Statement of Retained Earnings

Answer: b. Cash Flow Statement
Explanation: The Cash Flow Statement tracks the net change in cash resulting from various activities, providing a comprehensive view of cash inflows and outflows over a period.

259. When creating a budget for a construction project, which of the following is the FIRST step?
a. Allocating funds for unexpected expenses
b. Estimating labor costs
c. Determining the project's scope and objectives
d. Calculating material costs

Answer: c. Determining the project's scope and objectives
Explanation: Before diving into specifics, it's crucial to understand the project's scope and objectives. This foundational step informs all subsequent budgeting decisions.

260. Which of the following is NOT typically included in a construction project's direct costs?
a. Equipment rental fees
b. Building permits
c. Marketing and advertising
d. Raw materials

Answer: c. Marketing and advertising
Explanation: Direct costs are expenses directly tied to the production of a specific project. Marketing and advertising are considered indirect costs.

261. A construction contractor is concerned about potential cost overruns. Which budgeting technique allows for a contingency or buffer?
a. Zero-based budgeting
b. Fixed budgeting
c. Incremental budgeting
d. Flexible budgeting

Answer: d. Flexible budgeting. Explanation: Flexible budgeting allows for adjustments based on actual project needs or unexpected expenses, providing a contingency for unforeseen costs.

262. Which tax is directly related to the wages a construction company pays its employees?
a. Income tax
b. Sales tax
c. Payroll tax
d. Property tax

Answer: c. Payroll tax. Explanation: Payroll taxes are levied on wages and salaries and are used primarily to fund social insurance programs.

263. A construction contractor who sells building materials would most likely be concerned with which type of tax?
a. Income tax
b. Sales tax
c. Payroll tax
d. Excise tax

Answer: b. Sales tax
Explanation: Sales tax is imposed on the sale of goods and services. A contractor selling building materials would collect and remit this tax.

264. In which budgeting method does every expense need to be justified for each new period, starting from a "zero base"?
a. Zero-based budgeting
b. Fixed budgeting
c. Incremental budgeting
d. Flexible budgeting

Answer: a. Zero-based budgeting
Explanation: Zero-based budgeting requires every function within an organization to be analyzed and justified from the ground up, starting from zero.

265. Which of the following is a potential consequence for a construction contractor failing to meet tax obligations?
a. Increased sales
b. Tax liens on business properties
c. Decreased labor costs
d. Enhanced business reputation

Answer: b. Tax liens on business properties
Explanation: If a contractor fails to meet tax obligations, the government can place a lien on business properties until the debt is settled.

266. When considering indirect costs for a construction project budget, which of the following would be relevant?
a. Cost of bricks and mortar
b. Wages for construction workers on the project
c. Office administrative salaries
d. Cost of specific construction permits

Answer: c. Office administrative salaries
Explanation: Indirect costs are not directly tied to a specific project. Administrative salaries, which support the broader business operations, are considered indirect costs.

267. A construction company is expanding its operations and needs to account for increased overhead costs in its budget. Which of the following would be considered an overhead cost?
a. Cost of construction machinery
b. Wages for a specific project's labor
c. Rent for the company's main office
d. Cost of raw materials for a project

Answer: c. Rent for the company's main office
Explanation: Overhead costs are ongoing business expenses not directly tied to a specific project. Rent for the company's main office is an example of an overhead cost.

268. Which of the following is a primary reason for a construction company to maintain a regular and detailed financial analysis?
a. To ensure compliance with building codes
b. To monitor and manage cash flow effectively
c. To reduce the need for insurance
d. To decrease the company's tax obligations

Answer: b. To monitor and manage cash flow effectively
Explanation: Regular financial analysis allows a company to understand its financial health, ensuring that it can meet its obligations and invest in future opportunities.

269. A construction company is experiencing consistent delays in receiving payments from clients. Which of the following strategies would be MOST effective in improving cash flow?
a. Reducing overhead costs
b. Offering early payment discounts
c. Increasing project estimates
d. Delaying payments to suppliers

Answer: b. Offering early payment discounts
Explanation: Offering early payment discounts can incentivize clients to pay sooner, thus improving cash flow.

270. Which of the following is a primary reason for cost overruns in construction projects?
a. Overestimation of initial project costs
b. Efficient project management
c. Unexpected site conditions or design changes
d. Decreased labor costs

Answer: c. Unexpected site conditions or design changes
Explanation: Unforeseen site conditions or changes in design can lead to increased costs that were not initially budgeted for.

271. A contractor is considering taking a loan to finance a large project. Which of the following would be the LEAST important factor to consider?
a. The interest rate of the loan
b. The company's current debt levels
c. The color of the bank's logo
d. The loan's repayment terms

Answer: c. The color of the bank's logo
Explanation: The color of the bank's logo has no bearing on the financial implications of the loan.

272. Which financial statement provides a snapshot of a construction company's assets, liabilities, and equity at a specific point in time?
a. Income statement
b. Cash flow statement
c. Balance sheet
d. Statement of retained earnings

Answer: c. Balance sheet. Explanation: The balance sheet provides a snapshot of a company's financial position, detailing assets, liabilities, and equity.

273. For a construction company, which of the following would be considered a variable cost?
a. Rent for office space
b. Salaries of permanent staff
c. Cost of construction materials
d. Insurance premiums

Answer: c. Cost of construction materials. Explanation: The cost of construction materials can vary based on the scope and requirements of each project, making it a variable cost.

274. Which of the following is a key component of effective cost control in construction projects?
a. Ignoring small expenses as they are insignificant
b. Regularly reviewing and comparing budgeted vs. actual costs
c. Only focusing on labor costs
d. Avoiding the use of technology in cost estimation

Answer: b. Regularly reviewing and comparing budgeted vs. actual costs
Explanation: Regular reviews allow for timely adjustments and ensure that the project stays within the budget.

275. A construction company's revenue has been steadily increasing, but profits have been declining. What could be a potential reason?
a. Decreased competition in the market
b. Improved project management techniques
c. Rising operational costs
d. Reduced marketing expenses

Answer: c. Rising operational costs
Explanation: Even if revenue is increasing, rising operational costs can erode profits.

276. In the context of financial planning for a construction company, what is the primary purpose of a contingency fund?
a. To fund company outings and events
b. To invest in stocks and bonds
c. To cover unexpected expenses or emergencies
d. To pay regular operational costs

Answer: c. To cover unexpected expenses or emergencies
Explanation: A contingency fund acts as a financial safety net, ensuring the company can handle unforeseen costs without jeopardizing operations.

277. Which of the following would be considered a long-term liability for a construction company?
a. Outstanding invoices from suppliers
b. A mortgage on the company's office building
c. Wages payable to workers
d. Monthly utility bills

Answer: b. A mortgage on the company's office building
Explanation: A mortgage is typically repaid over many years, making it a long-term liability.

278. To enhance revenue, a construction company is considering diversifying its services. Which of the following would be the MOST important factor to consider before making this decision?
a. The company's current brand color
b. Market demand for the new service
c. The CEO's personal preferences
d. The number of employees in the company

Answer: b. Market demand for the new service
Explanation: Before diversifying, it's crucial to assess whether there's sufficient market demand to justify the expansion.

279. A construction company is reviewing its income statement and notices a consistent decline in net income. Which of the following would be the BEST initial step in risk management?
a. Ignore the trend and focus on increasing sales.
b. Seek a loan to cover any potential financial shortfalls.
c. Analyze the statement to identify areas of increased expenses.
d. Reduce the workforce to save on labor costs.

Answer: c. Analyze the statement to identify areas of increased expenses.
Explanation: Identifying areas of increased expenses can help address the root cause of declining net income and manage financial risks.

280. When budgeting for a construction project, why is it crucial to allocate a contingency fund?
a. To cover the cost of company events.
b. To invest in new business opportunities.
c. To manage financial risks associated with unforeseen project costs.
d. To pay regular operational costs.

Answer: c. To manage financial risks associated with unforeseen project costs.
Explanation: A contingency fund acts as a financial safety net for unexpected costs, helping to manage financial risks.

281. Which business structure in the construction industry has the potential for double taxation?
a. Sole Proprietorship
b. Partnership
c. Corporation
d. Limited Liability Company

Answer: c. Corporation
Explanation: Corporations can face double taxation, where the company's profits are taxed, and then shareholders are taxed on dividends.

282. A sole proprietorship in the construction industry is facing financial challenges. Which of the following is a risk specific to this business structure?
a. The business's debts are separate from the owner's personal debts.
b. The owner has limited liability for business debts.
c. The owner has unlimited personal liability for business debts.
d. The business has a perpetual existence.

Answer: c. The owner has unlimited personal liability for business debts.
Explanation: In a sole proprietorship, the owner is personally liable for all business debts, which can pose a significant financial risk.

283. How can a financial advisor assist a construction company in managing financial risks?
a. By taking over all decision-making responsibilities.
b. By providing guidance on investment opportunities.
c. By designing the company's logo.
d. By recommending new construction techniques.

Answer: b. By providing guidance on investment opportunities.
Explanation: Financial advisors can offer insights into viable investment opportunities, helping the company grow and manage financial risks.

284. Which of the following financial statements provides insights into a construction company's ability to generate cash to meet its financial obligations?
a. Income Statement
b. Statement of Retained Earnings
c. Balance Sheet
d. Cash Flow Statement

Answer: d. Cash Flow Statement
Explanation: The Cash Flow Statement provides detailed information about a company's cash inflows and outflows, indicating its ability to meet financial obligations.

285. For a construction corporation, which of the following is a primary consideration in financial risk management?
a. Protecting the personal assets of shareholders.
b. Ensuring double taxation is applied.
c. Avoiding the need for financial statements.
d. Ensuring all partners agree on financial decisions.

Answer: a. Protecting the personal assets of shareholders.
Explanation: One of the advantages of a corporation is the separation of business and personal assets, which is crucial for financial risk management.

286. A construction company is considering expanding its operations. Which financial statement would be MOST useful in determining the company's ability to finance this expansion internally?
a. Income Statement
b. Cash Flow Statement
c. Balance Sheet
d. Statement of Owner's Equity

Answer: b. Cash Flow Statement
Explanation: The Cash Flow Statement provides insights into the company's liquidity and its ability to finance operations without external funding.

287. In the context of financial risk management, why is it essential for construction companies to maintain accurate and up-to-date financial statements?
a. To ensure compliance with local construction regulations.
b. To provide data for marketing materials.
c. To make informed financial decisions and identify potential risks.
d. To determine the company's market share.

Answer: c. To make informed financial decisions and identify potential risks.
Explanation: Accurate financial statements allow companies to assess their financial health, make informed decisions, and identify areas of risk.

288. A construction company has hired an accountant for the first time. How can the accountant assist in financial risk management?
a. By taking over the company's marketing strategies.
b. By ensuring accurate financial reporting and compliance with tax obligations.
c. By managing the company's construction projects.
d. By negotiating contracts with clients.

Answer: b. By ensuring accurate financial reporting and compliance with tax obligations.
Explanation: An accountant can ensure that the company's financial records are accurate, which is crucial for assessing financial health and managing risks. Additionally, compliance with tax obligations can prevent costly penalties and legal issues.

289. A contractor is working on a historic building renovation. Which of the following would be the MOST important consideration regarding building codes?
a. Modern energy efficiency standards.
b. Current seismic retrofitting requirements.
c. Historic preservation guidelines.
d. Contemporary plumbing standards.

Answer: c. Historic preservation guidelines.
Explanation: While all building codes are essential, when working on historic buildings, preservation guidelines often take precedence to maintain the building's historic character.

290. In a region prone to earthquakes, which building code provision would be MOST relevant?
a. Snow load requirements.
b. Seismic design criteria.
c. Hurricane wind resistance.
d. Floodplain construction standards.

Answer: b. Seismic design criteria.
Explanation: In earthquake-prone areas, seismic design criteria are crucial to ensure the structural integrity of buildings during seismic events.

291. A contractor is building a commercial property in a coastal area. Which of the following regulations would be LEAST relevant?
a. Tsunami evacuation routes.
b. Saltwater corrosion resistance.
c. Tornado shelter requirements.
d. Flood-resistant construction.

Answer: c. Tornado shelter requirements.
Explanation: While tornadoes can occur in many places, coastal areas are more concerned with issues like tsunamis, corrosion from saltwater, and flooding.

292. Which of the following would be the primary reason for a city to update its building codes?
a. To increase city revenue from permit fees.
b. To reflect the latest safety and construction standards.
c. To reduce the number of buildings in the city.
d. To promote a specific architectural style.

Answer: b. To reflect the latest safety and construction standards.
Explanation: Building codes are primarily updated to incorporate the latest in safety, technology, and construction practices.

293. A contractor is working on a high-rise project in a densely populated city. Which of the following building code considerations would be LEAST relevant?
a. Elevator safety standards.
b. Rural septic system guidelines.
c. Fire evacuation procedures.
d. Structural load-bearing standards.

Answer: b. Rural septic system guidelines.
Explanation: In a densely populated city, especially for a high-rise, rural septic systems would not be applicable.

294. Which organization is primarily responsible for developing model building codes used by many U.S. municipalities?
a. American Society of Civil Engineers (ASCE).
b. Occupational Safety and Health Administration (OSHA).
c. International Code Council (ICC).
d. National Association of Home Builders (NAHB).

Answer: c. International Code Council (ICC).
Explanation: The ICC develops the International Building Code (IBC) and other model codes used by many municipalities.

295. A contractor is unsure about a specific building code requirement. What should be their FIRST step?
a. Proceed as they think best and correct it later if needed.
b. Ask a fellow contractor for their opinion.
c. Consult the local building department or the specific code in question.
d. Base the decision on what was done in past projects.

Answer: c. Consult the local building department or the specific code in question.
Explanation: The local building department or the specific code provides the definitive answer on code requirements.

296. In terms of building codes, what is the primary purpose of a "setback" regulation?
a. To ensure buildings are energy efficient.
b. To determine the distance a building must be from the property line or another structure.
c. To establish the height limit of a structure.
d. To ensure buildings have a modern aesthetic.

Answer: b. To determine the distance a building must be from the property line or another structure.
Explanation: Setback regulations define how far structures must be set back from property lines, roads, or other structures.

297. A contractor discovers that a completed project does not comply with a specific building code. What is the MOST likely consequence?
a. The building will be immediately demolished.
b. The contractor will receive a financial bonus.
c. The contractor will need to modify the project to achieve compliance.
d. The building code will be changed to match the project.

Answer: c. The contractor will need to modify the project to achieve compliance.
Explanation: If a project does not meet code, the contractor will typically need to bring it into compliance, which may involve modifications or repairs.

298. Which of the following would NOT typically be covered under building codes?
a. Structural integrity of a building.
b. Energy efficiency standards.
c. Interior design aesthetics.
d. Electrical system safety.

Answer: c. Interior design aesthetics.
Explanation: While some codes might touch on aspects of design (like historic preservation), general interior design aesthetics are not the primary focus of building codes.

299. The International Building Code (IBC) primarily serves to:
a. Promote a specific architectural style.
b. Ensure public health, safety, and welfare in the built environment.
c. Increase city revenue from permit fees.
d. Encourage innovative design.

Answer: b. Ensure public health, safety, and welfare in the built environment.
Explanation: The IBC's primary purpose is to ensure the safety and welfare of the public in relation to the built environment.

300. Which of the following is NOT a primary focus of the IBC?
a. Structural integrity.
b. Fire safety.
c. Energy efficiency.
d. Color schemes of exterior walls.

Answer: d. Color schemes of exterior walls.
Explanation: The IBC focuses on safety and structural standards, not aesthetic choices like color schemes.

301. When considering variations in building codes between trades, which trade would be MOST concerned with vent stack configurations?
a. Electrical.
b. Plumbing.
c. HVAC.
d. Carpentry.

Answer: b. Plumbing.
Explanation: Vent stack configurations are a primary concern in plumbing to ensure proper drainage and prevent sewer gas from entering buildings.

302. A contractor is working on an HVAC system in a commercial building. Which code would they primarily refer to for standards and regulations?
a. National Electrical Code (NEC).
b. International Plumbing Code (IPC).
c. International Mechanical Code (IMC).
d. International Fire Code (IFC).

Answer: c. International Mechanical Code (IMC).
Explanation: The IMC provides standards for HVAC and mechanical systems in buildings.

303. Before starting a major renovation project, a contractor should FIRST:
a. Begin demolition.
b. Secure financing.
c. Obtain necessary construction permits.
d. Advertise the project to the public.

Answer: c. Obtain necessary construction permits. Explanation: Before starting work, it's crucial to obtain the necessary permits to ensure the project complies with local regulations.

304. Which of the following would be a likely consequence of failing an inspection due to not meeting code requirements?
a. The contractor will be immediately licensed for additional trades.
b. The project can continue without any changes.
c. The contractor will need to modify the work to meet code before proceeding.
d. The inspector will be reassigned.

Answer: c. The contractor will need to modify the work to meet code before proceeding.
Explanation: If a project fails an inspection, the contractor typically must bring the work into compliance with the code before continuing.

305. In the context of electrical installations, which code provides comprehensive regulations to ensure electrical safety?
a. International Residential Code (IRC).
b. International Fire Code (IFC).
c. National Electrical Code (NEC).
d. International Energy Conservation Code (IECC).

Answer: c. National Electrical Code (NEC). Explanation: The NEC, also known as NFPA 70, is the standard for electrical safety in residential, commercial, and industrial settings.

306. A contractor discovers that a completed project does not comply with a specific building code. What is the MOST likely immediate consequence?
a. The building will be immediately demolished.
b. The contractor will receive a financial bonus.
c. The contractor will be issued a "stop work" order or a violation notice.
d. The building code will be changed to match the project.

Answer: c. The contractor will be issued a "stop work" order or a violation notice.
Explanation: If a project is found not to meet code, the local building department may issue a "stop work" order or a violation notice, requiring the contractor to address the issue.

307. Which of the following is NOT a primary reason for a city to require construction permits?
a. To ensure public safety by enforcing building codes.
b. To generate revenue for the city.
c. To keep a record of all construction in the city.
d. To give preference to certain contractors.

Answer: d. To give preference to certain contractors.
Explanation: Construction permits are not meant to show favoritism but to ensure safety, generate revenue, and maintain records.

308. When considering the IBC and its impact on construction standards, which of the following would be the LEAST likely focus of the code?
a. Ensuring buildings can withstand natural disasters.
b. Making sure buildings are accessible to people with disabilities.
c. Promoting energy efficiency in new constructions.
d. Dictating the artistic style of building interiors.

Answer: d. Dictating the artistic style of building interiors.
Explanation: While the IBC addresses safety, accessibility, and efficiency, it doesn't dictate artistic or aesthetic choices.

309. Per the IBC 2018, which section specifically addresses the requirements for fire-resistant construction in various types of buildings?
a. Section 701
b. Section 1001
c. Section 1502
d. Section 2603

Answer: a. Section 701
Explanation: Section 701 of the IBC 2018 specifically deals with the general requirements for fire-resistant construction.

310. According to the NEC (NFPA 70), which article provides the general requirements for electrical installations?
a. Article 90
b. Article 100
c. Article 110
d. Article 200

Answer: c. Article 110
Explanation: Article 110 of the NEC provides the general requirements for electrical installations.

311. In the International Mechanical Code (IMC) 2018, which section provides the general regulations for mechanical systems?
a. Section 101
b. Section 202
c. Section 301
d. Section 401

Answer: a. Section 101
Explanation: Section 101 of the IMC 2018 provides the scope and general regulations for mechanical systems.

312. According to the International Plumbing Code (IPC) 2018, which section addresses the general regulations for plumbing systems?
a. Section 101
b. Section 202
c. Section 301
d. Section 401

Answer: c. Section 301
Explanation: Section 301 of the IPC 2018 provides the general regulations for plumbing systems.

313. Per the IBC 2018, which section specifically addresses the requirements for accessibility in building design and construction?
a. Section 1001
b. Section 1009
c. Section 1101
d. Section 1201

Answer: c. Section 1101
Explanation: Section 1101 of the IBC 2018 specifically deals with the general requirements for accessibility in building design and construction.

314. Which of the following is a widely recognized green building certification system?
a. EnergyPlus
b. LEED
c. BREEAM
d. ASHRAE 90.1

Answer: b. LEED
Explanation: LEED (Leadership in Energy and Environmental Design) is a globally recognized green building certification system.

315. The International Energy Conservation Code (IECC) primarily focuses on:
a. Water conservation
b. Waste management
c. Energy efficiency in buildings
d. Renewable energy sources

Answer: c. Energy efficiency in buildings
Explanation: The IECC provides building codes specifically designed to promote energy-efficient buildings.

316. Which of the following is NOT a primary goal of green building standards?
a. Reducing water usage
b. Enhancing indoor environmental quality
c. Increasing construction costs
d. Reducing energy consumption

Answer: c. Increasing construction costs
Explanation: Green building standards aim to make construction more sustainable and efficient, not to increase costs.

317. ASHRAE 90.1 is a standard that provides:
a. Guidelines for indoor air quality
b. Minimum requirements for energy-efficient building design
c. Water conservation techniques
d. Waste management procedures

Answer: b. Minimum requirements for energy-efficient building design
Explanation: ASHRAE 90.1 sets the minimum energy-saving requirements for building design.

318. Net-zero energy buildings aim to:
a. Use no water
b. Produce as much energy as they consume
c. Have zero carbon emissions
d. Use no artificial lighting

Answer: b. Produce as much energy as they consume
Explanation: Net-zero energy buildings are designed to produce as much energy on-site as they consume on an annual basis.

319. Which organization developed the Green Globes assessment and rating system?
a. U.S. Green Building Council
b. Green Building Initiative
c. International Code Council
d. American Society of Heating, Refrigerating, and Air-Conditioning Engineers

Answer: b. Green Building Initiative
Explanation: The Green Building Initiative developed the Green Globes system as an alternative to LEED.

320. Passive design strategies in green building focus on:
a. Active mechanical systems
b. Utilizing natural energy sources without active systems
c. Implementing renewable energy systems
d. Using high-tech building management systems

Answer: b. Utilizing natural energy sources without active systems
Explanation: Passive design uses natural sources of heating, cooling, and ventilation without relying on mechanical systems.

321. Which of the following is a benefit of green roofs?
a. Increase in HVAC load
b. Reduction in stormwater runoff
c. Increase in building's heat island effect
d. Reduction in indoor air quality

Answer: b. Reduction in stormwater runoff
Explanation: Green roofs can absorb rainwater, reducing stormwater runoff and providing insulation to reduce the building's energy needs.

322. Envelope commissioning in green building primarily focuses on:
a. HVAC systems
b. Building's exterior shell
c. Lighting systems
d. Water systems

Answer: b. Building's exterior shell
Explanation: Envelope commissioning focuses on the building's exterior shell, ensuring it performs efficiently to reduce energy consumption.

323. Which of the following is NOT a primary focus of the WELL Building Standard?
a. Nutrition
b. Light
c. Acoustics
d. Energy efficiency

Answer: d. Energy efficiency
Explanation: The WELL Building Standard focuses on human health and well-being, addressing aspects like air, water, nourishment, light, and more. Energy efficiency, while important, is not its primary focus.

324. Which section of the ADA Standards for Accessible Design specifically addresses accessible routes within buildings?
a. Section 206
b. Section 309
c. Section 502
d. Section 605

Answer: a. Section 206
Explanation: Section 206 of the ADA Standards for Accessible Design deals with accessible routes, detailing requirements for paths that allow people with disabilities to access all usable spaces within a building.

325. The International Building Code (IBC) requires that exit access doorways be spaced at least how far apart, measured from door center to door center, in most situations?
a. One-fourth the length of the maximum overall diagonal dimension of the building area
b. One-third the length of the maximum overall diagonal dimension of the building area
c. One-half the length of the maximum overall diagonal dimension of the building area
d. Equal to the length of the maximum overall diagonal dimension of the building area

Answer: c. One-half the length of the maximum overall diagonal dimension of the building area
Explanation: According to the IBC, exit access doorways should be spaced at least half the length of the maximum overall diagonal dimension of the building area to ensure safety during evacuations.

326. For a commercial building, which of the following is NOT a primary concern related to ADA compliance?
a. Elevator size
b. Color of the exterior paint
c. Width of hallways
d. Ramp gradients

Answer: b. Color of the exterior paint
Explanation: The ADA does not specify requirements related to the color of exterior paint. It focuses on ensuring accessibility features, such as elevators, hallways, and ramps, meet specific standards.

327. Which NFPA standard is primarily concerned with the installation of sprinkler systems?
a. NFPA 70
b. NFPA 72
c. NFPA 13
d. NFPA 101

Answer: c. NFPA 13
Explanation: NFPA 13 is the standard for the installation of sprinkler systems, ensuring they are installed to effectively control and suppress fires.

328. In terms of structural safety, the live load of a building refers to:
a. The weight of the building's walls and roof
b. The weight of permanent installations like HVAC systems
c. The weight of temporary occupants and movable objects
d. The weight of the building's foundation

Answer: c. The weight of temporary occupants and movable objects
Explanation: Live loads are temporary loads that the building must support, such as occupants, furniture, and snow.

329. Which of the following would NOT typically be a consideration under ADA requirements for parking?
a. Number of accessible parking spaces
b. Size of each parking space
c. Color of parking space markings
d. Location of accessible parking spaces relative to the building entrance

Answer: c. Color of parking space markings
Explanation: While the ADA provides guidelines on the number, size, and location of accessible parking spaces, it doesn't specify the color of parking space markings.

330. Which code provides guidelines on fire-resistance-rated construction and fire protection systems in buildings?
a. IBC (International Building Code)
b. IRC (International Residential Code)
c. IMC (International Mechanical Code)
d. IPC (International Plumbing Code)

Answer: a. IBC (International Building Code)
Explanation: The IBC provides guidelines on various aspects of building construction, including fire-resistance-rated construction and fire protection systems.

331. For a multi-story office building, which of the following would be the primary concern related to ADA compliance?
a. The height of the building
b. The number of windows on each floor
c. Elevator accessibility and operation
d. The type of roofing material used

Answer: c. Elevator accessibility and operation
Explanation: For multi-story buildings, ensuring that all floors are accessible is crucial. Elevator accessibility and operation are primary concerns under ADA guidelines.

332. Which of the following is a primary structural safety concern when constructing buildings in earthquake-prone areas?
a. Ensuring flexible building connections
b. Using heavier building materials
c. Increasing the number of windows
d. Using rigid building designs

Answer: a. Ensuring flexible building connections
Explanation: In earthquake-prone areas, buildings need to have some flexibility to withstand seismic forces. Flexible building connections help in absorbing and dissipating energy during an earthquake.

333. Which NFPA standard is known as the "Life Safety Code" and addresses building construction, protection, and occupancy features?
a. NFPA 70
b. NFPA 72
c. NFPA 13
d. NFPA 101

Answer: d. NFPA 101
Explanation: NFPA 101, known as the "Life Safety Code," provides guidelines to ensure the safety of occupants in buildings and structures.

334. Which section of the International Building Code (IBC) specifically addresses the structural design criteria for buildings?
a. Chapter 5
b. Chapter 16
c. Chapter 10
d. Chapter 21

Answer: b. Chapter 16
Explanation: Chapter 16 of the IBC provides guidelines on the structural design of buildings, ensuring they meet certain criteria for safety and stability.

335. The National Electrical Code (NEC) is also known by which NFPA standard number?
a. NFPA 70
b. NFPA 72
c. NFPA 13
d. NFPA 101

Answer: a. NFPA 70
Explanation: The National Electrical Code (NEC) is also known as NFPA 70 and provides guidelines for electrical installations.

336. Which of the following is NOT a primary focus of the International Residential Code (IRC)?
a. Electrical systems in residential buildings
b. Plumbing systems in commercial buildings
c. Mechanical systems in residential buildings
d. Building and energy conservation in residential buildings

Answer: b. Plumbing systems in commercial buildings
Explanation: The IRC primarily focuses on residential buildings, so plumbing systems in commercial buildings are not its primary concern.

337. In terms of fire safety, which section of the IBC discusses fire-resistant materials and construction?
a. Chapter 7
b. Chapter 10
c. Chapter 16
d. Chapter 21

Answer: a. Chapter 7
Explanation: Chapter 7 of the IBC provides guidelines on fire-resistant materials and construction techniques to ensure building safety.

338. Which of the following would be a primary concern under the International Plumbing Code (IPC)?
a. Electrical circuit design
b. Sanitary drainage systems
c. Building insulation
d. Fire sprinkler systems

Answer: b. Sanitary drainage systems
Explanation: The IPC provides guidelines on plumbing systems, including sanitary drainage systems.

339. The International Mechanical Code (IMC) primarily addresses which of the following?
a. Structural design of buildings
b. HVAC and refrigeration systems
c. Electrical installations
d. Fire safety measures

Answer: b. HVAC and refrigeration systems
Explanation: The IMC provides guidelines for the design, installation, and maintenance of HVAC and refrigeration systems.

340. Which code would a contractor refer to for guidelines on energy efficiency in buildings?
a. International Energy Conservation Code (IECC)
b. International Fire Code (IFC)
c. International Residential Code (IRC)
d. National Electrical Code (NEC)

Answer: a. International Energy Conservation Code (IECC)
Explanation: The IECC provides guidelines on energy conservation and efficiency in building design and construction.

341. For a commercial building, which of the following codes would be most relevant for ensuring accessibility for individuals with disabilities?
a. ADA Standards for Accessible Design
b. International Fire Code (IFC)
c. National Electrical Code (NEC)
d. International Mechanical Code (IMC)

Answer: a. ADA Standards for Accessible Design
Explanation: The ADA Standards for Accessible Design provide guidelines to ensure buildings and facilities are accessible to individuals with disabilities.

342. Which of the following is NOT a primary focus of the International Green Construction Code (IgCC)?
a. Energy efficiency
b. Water conservation
c. Sustainable construction materials
d. Electrical circuit design

Answer: d. Electrical circuit design
Explanation: While the IgCC covers various aspects of green and sustainable construction, electrical circuit design is not its primary focus.

343. In terms of structural safety, which code provides guidelines for the design and construction of buildings in seismic zones?
a. International Seismic Code (ISC)
b. International Building Code (IBC)
c. National Electrical Code (NEC)
d. International Fire Code (IFC)

Answer: b. International Building Code (IBC)
Explanation: The IBC provides guidelines for building design and construction in seismic zones to ensure structural safety during earthquakes.

344. Which of the following materials is best suited for exterior cladding due to its durability and resistance to environmental factors?
a. Softwood lumber
b. Gypsum board
c. Fiber-cement siding
d. Particleboard

Answer: c. Fiber-cement siding
Explanation: Fiber-cement siding is known for its durability, resistance to rot, pests, and weather conditions, making it ideal for exterior cladding.

345. When installing a suspended ceiling in a commercial building, which tool would be essential for ensuring the grid is level?
a. Plumb bob
b. Claw hammer
c. Laser level
d. Pipe wrench

Answer: c. Laser level
Explanation: A laser level projects a perfectly straight line on surfaces, ensuring that the suspended ceiling grid is level across the entire space.

346. Which material is commonly used for its fire-resistant properties in commercial building walls?
a. Plywood
b. Type X gypsum board
c. Hardwood
d. MDF board

Answer: b. Type X gypsum board
Explanation: Type X gypsum board is specially designed with fire-resistant properties and is commonly used in commercial construction for walls and ceilings.

347. For a contractor looking to achieve a polished concrete floor, which of the following methods would be most appropriate?
a. Sandblasting
b. Diamond grinding
c. Power troweling
d. Shotcreting

Answer: b. Diamond grinding
Explanation: Diamond grinding uses diamond-tipped blades to smooth and polish concrete surfaces, making it the ideal method for achieving a polished concrete floor.

348. Which tool is essential for a contractor installing ceramic tiles to ensure even spacing between tiles?
a. Tile nipper
b. Tile spacers
c. Notched trowel
d. Tile cutter

Answer: b. Tile spacers
Explanation: Tile spacers are placed between tiles during installation to ensure consistent and even gaps for grout.

349. In the context of roofing, which material is known for its lightweight properties and is often used for its reflective capabilities?
a. Asphalt shingles
b. Clay tiles
c. Metal roofing
d. Slate tiles

Answer: c. Metal roofing
Explanation: Metal roofing is lightweight and can be treated to be highly reflective, reducing energy costs by reflecting sunlight.

350. Which method is commonly used to join copper pipes in plumbing systems?
a. Welding
b. Soldering
c. Bolting
d. Gluing

Answer: b. Soldering
Explanation: Soldering is the process of joining two metal items by melting and flowing a filler metal into the joint, and it's commonly used for copper pipes in plumbing.

351. For a contractor working with masonry, which tool is essential for ensuring the bricks are laid level and in a straight line?
a. Mason's chisel
b. Tuck pointer
c. Mason's line
d. Brick hammer

Answer: c. Mason's line
Explanation: A mason's line is stretched between two points to provide a straight and level guide for laying bricks or blocks.

352. Which of the following materials is known for its insulating properties and is often sprayed into wall cavities?
a. Plywood
b. Spray foam insulation
c. Particleboard
d. Cement board

Answer: b. Spray foam insulation
Explanation: Spray foam insulation expands upon application, filling wall cavities and providing excellent thermal insulation.

353. In terms of flooring, which method involves installing wood flooring planks at a 45-degree angle to the walls?
a. Parquet
b. Herringbone
c. Diagonal
d. Straight lay

Answer: c. Diagonal
Explanation: Installing wood flooring planks at a 45-degree angle to the walls is known as the diagonal method, which can add visual interest to a room.

354. Which material is known for its high tensile strength and is commonly used as reinforcement in concrete structures?
a. Aluminum
b. Copper
c. Steel rebar
d. PVC

Answer: c. Steel rebar
Explanation: Steel rebar (reinforcing bar) is embedded in concrete to provide tensile strength, allowing the concrete to withstand stretching and bending forces.

355. When considering thermal insulation properties for a residential building, which material offers the highest R-value per inch?
a. Fiberglass batts
b. Expanded polystyrene (EPS)
c. Closed-cell spray foam
d. Cellulose

Answer: c. Closed-cell spray foam
Explanation: Closed-cell spray foam typically offers a higher R-value per inch compared to other insulation materials, making it highly effective in insulating spaces.

356. For exterior cladding in areas prone to wildfires, which material is considered non-combustible and offers the best fire resistance?
a. Vinyl siding
b. Wood cladding
c. Stucco
d. Cedar shingles

Answer: c. Stucco
Explanation: Stucco is a non-combustible material that offers excellent fire resistance, making it suitable for areas with high wildfire risks.

357. Which material is preferred for its soundproofing capabilities in partition walls of commercial buildings?
a. Plywood
b. Drywall
c. Acoustic panels
d. Hardboard

Answer: c. Acoustic panels
Explanation: Acoustic panels are specifically designed to absorb sound and reduce noise transmission, making them ideal for soundproofing applications.

358. In roofing, which material is known to have a lifespan of over 100 years when properly maintained?
a. Asphalt shingles
b. Slate tiles
c. Metal roofing
d. Wood shakes

Answer: b. Slate tiles
Explanation: Slate tiles are incredibly durable and can last for over a century with proper maintenance, outlasting many other roofing materials.

359. For a high-traffic commercial space, which flooring material is renowned for its durability and ease of maintenance?
a. Carpet
b. Terrazzo
c. Laminate
d. Parquet

Answer: b. Terrazzo
Explanation: Terrazzo is a composite material that is both durable and easy to maintain, making it ideal for high-traffic areas.

360. Which material is commonly used in window frames for its thermal efficiency and low maintenance requirements?
a. Aluminum
b. Steel
c. Vinyl
d. Wood

Answer: c. Vinyl
Explanation: Vinyl window frames are known for their thermal efficiency, resistance to weathering, and low maintenance needs.

361. In terms of sustainability, which material is considered renewable and is often used in green building projects?
a. PVC
b. Concrete
c. Bamboo
d. Fiberglass

Answer: c. Bamboo
Explanation: Bamboo grows rapidly and is considered a renewable resource, making it a popular choice for sustainable construction projects.

362. For underground water supply lines, which material is preferred due to its resistance to corrosion and longevity?
a. Galvanized steel
b. Copper
c. PVC
d. Cast iron

Answer: b. Copper
Explanation: Copper is resistant to corrosion and is commonly used for underground water supply lines due to its longevity and reliability.

363. Which material, when used in exterior walls, provides both insulation and structural support?
a. Drywall
b. Insulated concrete forms (ICFs)
c. Plywood sheathing
d. Vapor barrier

Answer: b. Insulated concrete forms (ICFs)
Explanation: ICFs are a system of formwork for reinforced concrete that stays in place as a permanent interior and exterior substrate, providing both insulation and structural support.

364. Which tool is specifically designed to create joints in woodwork, allowing for pieces to be connected at right angles?
a. Doweling jig
b. Biscuit joiner
c. Brad nailer
d. Bench grinder

Answer: b. Biscuit joiner
Explanation: A biscuit joiner is used to cut crescent-shaped holes in the edges of wood. Wooden biscuits are then placed in these holes, and when two pieces are joined, they create a strong bond.

365. For a contractor aiming to measure the slope of a roof, which tool would be most appropriate?
a. Plumb bob
b. Framing square
c. Roofing calculator
d. Pitch gauge

Answer: d. Pitch gauge
Explanation: A pitch gauge is specifically designed to measure the slope or pitch of a roof, providing accurate readings for contractors.

366. When needing to create a hole in concrete without causing a fracture, which tool is most suitable?
a. Impact driver
b. Rotary hammer
c. Reciprocating saw
d. Angle grinder

Answer: b. Rotary hammer
Explanation: A rotary hammer is designed to drill into hard materials like concrete without causing fractures, making it ideal for such tasks.

367. Which tool is essential for electricians when they need to strip the insulation from wires?
a. Tin snips
b. Wire stripper
c. Bolt cutter
d. Lineman's pliers

Answer: b. Wire stripper
Explanation: A wire stripper is specifically designed to strip the insulation from electrical wires, making it a vital tool for electricians.

368. For a contractor laying tiles, which tool is crucial for ensuring even spacing between each tile?
a. Tile nipper
b. Tile cutter
c. Tile spacer
d. Grout float

Answer: c. Tile spacer
Explanation: Tile spacers are small pieces of plastic that ensure consistent spacing between tiles, ensuring a uniform appearance.

369. Which tool would a contractor use to check if a surface is perfectly horizontal or vertical?
a. Caliper
b. Spirit level
c. T-square
d. Protractor

Answer: b. Spirit level
Explanation: A spirit level contains a liquid-filled vial with a bubble that indicates whether a surface is level (horizontal) or plumb (vertical).

370. In HVAC work, which tool is essential for measuring the pressure of a refrigeration system?
a. Multimeter
b. Manifold gauge
c. Anemometer
d. Thermocouple

Answer: b. Manifold gauge
Explanation: A manifold gauge is used in HVAC to measure the pressure of refrigeration systems, helping technicians diagnose and service these systems.

371. For a contractor working with drywall, which tool is used to fasten the drywall to studs without using screws or nails?
a. Drywall saw
b. Drywall lifter
c. Drywall tape
d. Drywall adhesive applicator

Answer: d. Drywall adhesive applicator
Explanation: A drywall adhesive applicator is used to apply adhesive to fasten drywall sheets to studs without the need for screws or nails.

372. Which tool would be most appropriate for a contractor needing to cut a piece of glass for a window installation?
a. Glass cutter
b. Hacksaw
c. Jigsaw
d. Miter saw

Answer: a. Glass cutter
Explanation: A glass cutter is specifically designed to score glass, allowing it to be snapped cleanly along the scored line.

373. For a contractor working on a plumbing job, which tool is essential for connecting two pieces of copper pipe?
a. Pipe wrench
b. Flaring tool
c. Pipe cutter
d. Pipe bender

Answer: b. Flaring tool
Explanation: A flaring tool is used to widen the end of a copper pipe so it can be connected to another piece, creating a tight seal.

374. When constructing a load-bearing wall, which method is typically employed to distribute the weight of the structure above it?
a. Post-tensioning
b. Cantilevering
c. Cavity wall construction
d. Piling

Answer: d. Piling
Explanation: Piling involves driving piles into the ground below the structure to support and spread the load, ensuring stability for load-bearing walls.

375. In areas prone to earthquakes, which construction method is often used to allow buildings to withstand seismic activities?
a. Floating foundation
b. Base isolation
c. Slab-on-grade
d. Strip footing

Answer: b. Base isolation
Explanation: Base isolation involves placing isolators between the structure and its foundation, allowing the building to move independently of ground motion during an earthquake.

376. Which method is commonly used in the construction industry to achieve energy efficiency by minimizing thermal bridges in a building's envelope?
a. Cladding
b. SIPS (Structural Insulated Panels)
c. ICF (Insulated Concrete Forms)
d. Double skin façade

Answer: c. ICF (Insulated Concrete Forms)
Explanation: ICFs are formwork for concrete that stays in place as permanent building insulation for energy-efficient, cast-in-place, reinforced concrete walls.

377. For a contractor aiming to reduce noise transmission between rooms, which construction method would be most effective?
a. Single stud wall
b. Double stud wall
c. Staggered stud wall
d. Hollow block wall

Answer: b. Double stud wall
Explanation: Double stud walls provide an additional layer of separation, which can effectively reduce sound transmission between spaces.

378. Which method is typically employed in high-rise construction to ensure the verticality of the structure as it rises?
a. Slip forming
b. Tilt-up construction
c. Prefabrication
d. Modular construction

Answer: a. Slip forming
Explanation: Slip forming allows concrete to be poured continuously, ensuring that the structure remains vertical and consistent as it rises.

379. In green construction, which method is used to reduce stormwater runoff from a site?
a. Rain gardens
b. Vapor barriers
c. Silt fencing
d. Caissons

Answer: a. Rain gardens
Explanation: Rain gardens are designed to capture and absorb rainwater, reducing the amount of stormwater runoff from a site.

380. Which construction method is commonly used to prevent soil erosion in areas with steep slopes?
a. Gabion walls
b. Retaining walls
c. Parapet walls
d. Shear walls

Answer: b. Retaining walls
Explanation: Retaining walls are designed to hold back soil and prevent it from eroding, especially in areas with steep slopes.

381. For a contractor working on a bridge project, which method is employed to prevent the bridge from collapsing under dynamic loads, like wind or traffic?
a. Torsion reinforcement
b. Cantilevering
c. Damping
d. Bracing

Answer: c. Damping
Explanation: Damping systems are used in bridges to absorb and dissipate energy, preventing the bridge from resonating with dynamic loads.

382. In the construction of skyscrapers, which method is used to ensure the building can withstand lateral loads, such as wind?
a. Cross-bracing
b. Buttressing
c. Underpinning
d. Pile driving

Answer: a. Cross-bracing
Explanation: Cross-bracing involves using diagonal braces in the structure, which helps distribute lateral loads and provides stability against forces like wind.

383. Which method is commonly used in the construction industry to rapidly erect buildings by assembling factory-made components on-site?
a. Cast-in-place
b. Post-tensioning
c. Modular construction
d. Slip forming

Answer: c. Modular construction. Explanation: Modular construction involves creating entire sections of a building in a factory setting and then transporting and assembling them on-site, speeding up the construction process.

384. When installing a radiant floor heating system, which type of flooring material is considered the most effective in conducting and radiating heat?
a. Carpet
b. Vinyl
c. Tile
d. Laminate

Answer: c. Tile
Explanation: Tile is a good conductor of heat, making it effective for radiant floor heating systems. It retains heat well and radiates it evenly throughout the room.

385. For roofing in a high-wind area, which method is recommended to provide the best wind resistance?
a. Stapling shingles
b. Nailing shingles
c. Gluing shingles
d. Clipping shingles

Answer: b. Nailing shingles
Explanation: Nails, when properly installed, provide better wind resistance than staples. The number and placement of nails are crucial for optimal wind resistance.

386. In masonry, which type of mortar is best suited for load-bearing walls due to its high compressive strength?
a. Type N
b. Type O
c. Type S
d. Type M

Answer: d. Type M
Explanation: Type M mortar has the highest compressive strength, making it suitable for use in load-bearing walls and other structural applications.

387. For electrical installations in wet locations, which type of conduit is most appropriate?
a. EMT (Electrical Metallic Tubing)
b. PVC (Polyvinyl Chloride)
c. RMC (Rigid Metal Conduit)
d. FMC (Flexible Metal Conduit)

Answer: b. PVC (Polyvinyl Chloride)
Explanation: PVC conduit is resistant to moisture and corrosion, making it ideal for wet locations.

388. When installing plumbing in areas prone to freezing, which method is used to prevent pipes from bursting?
a. Using wider pipes
b. Installing pipe heaters
c. Pipe lagging
d. Using flexible PEX tubing

Answer: b. Installing pipe heaters
Explanation: Pipe heaters, or heat tape, can be wrapped around pipes to keep them warm and prevent freezing.

389. In HVAC, which method is considered more energy-efficient for cooling in dry climates?
a. Central air conditioning
b. Swamp coolers or evaporative coolers
c. Window units
d. Ductless mini-splits

Answer: b. Swamp coolers or evaporative coolers
Explanation: In dry climates, evaporative coolers are effective and use less energy than traditional air conditioners because they cool air through the evaporation of water.

390. For soundproofing between rooms in a commercial setting, which material is commonly used?
a. Polystyrene foam
b. Fiberglass insulation
c. Mass loaded vinyl
d. Polyurethane foam

Answer: c. Mass loaded vinyl
Explanation: Mass loaded vinyl is dense and effective at reducing sound transmission between spaces.

391. In carpentry, when joining two pieces of wood at right angles, which joint is considered the strongest?
a. Butt joint
b. Miter joint
c. Dovetail joint
d. Lap joint

Answer: c. Dovetail joint
Explanation: The dovetail joint, with its interlocking design, provides a large gluing area and mechanical strength, making it one of the strongest joints.

392. For exterior painting in humid climates, which type of paint is recommended due to its ability to resist mold and mildew?
a. Oil-based paint
b. Latex paint
c. Acrylic paint
d. Enamel paint

Answer: c. Acrylic paint
Explanation: Acrylic paint is water-resistant and less prone to support mold and mildew growth, making it suitable for humid climates.

393. In tiling, which type of adhesive is recommended for areas that will be frequently wet, like showers?
a. Mastic
b. Epoxy adhesive
c. Thin-set mortar
d. Liquid nails

Answer: c. Thin-set mortar
Explanation: Thin-set mortar is water-resistant and provides a strong bond, making it ideal for wet areas like showers.

394. Case 1: John, a contractor, is reviewing a set of blueprints for a new residential project. He notices a symbol that looks like a small triangle pointing to a line with numbers next to it.

What does this symbol most likely represent?
a. The direction of the prevailing wind.
b. The slope of the roof.
c. The location of a light fixture.
d. The elevation of the site.

Answer: b. The slope of the roof.
Explanation: In blueprints, a triangle pointing to a line with numbers often indicates the slope or pitch of the roof.

395. Case 2: Sarah is examining a blueprint of a commercial building. She sees a series of parallel lines on a section of the floor plan.
What do these parallel lines most likely indicate?
a. Stairs.
b. Elevator shaft.
c. Windows.
d. Wall section.

Answer: a. Stairs.
Explanation: Parallel lines on a floor plan typically represent stairs, with the number of lines corresponding to the number of steps.

396. Case 3: A contractor is reviewing a blueprint and notices a circular symbol with an "X" inside. This symbol is located in various rooms throughout the blueprint.
What does this symbol most likely represent?
a. Electrical outlets.
b. Ceiling fans.
c. Light fixtures.
d. Ventilation ducts.

Answer: a. Electrical outlets.
Explanation: The circular symbol with an "X" inside typically represents electrical outlets on blueprints.

397. Case 4: In a blueprint for a multi-story building, there's a note next to a door symbol that reads "36L."
What does "36L" typically indicate?
a. The door is 36 inches long.
b. The door opens to the left.
c. The door has a 36-inch lintel.
d. The door is located on the 36th floor.

Answer: b. The door opens to the left.
Explanation: In blueprints, door specifications like "36L" typically indicate the door's width (36 inches) and the direction it opens (Left).

398. Case 5: While examining a blueprint, a contractor notices a dashed line connecting two parallel solid lines. The dashed line has arrows on both ends.
What does this dashed line most likely represent?
a. A hidden structural beam.
b. The swing direction of a door.
c. An electrical circuit.
d. A ventilation duct.

Answer: b. The swing direction of a door.
Explanation: Dashed lines with arrows on blueprints typically indicate the swing direction of doors.

399. Case 6: A contractor is reviewing a blueprint for a kitchen remodel. He sees a symbol that looks like a circle divided into four quadrants.
What does this symbol most likely represent?
a. A ceiling light.
b. A floor drain.
c. An electrical outlet.
d. A sink.

Answer: d. A sink.
Explanation: In kitchen blueprints, a circle divided into four quadrants typically represents a sink.

400. Case 7: On a blueprint, there's a symbol that looks like a small rectangle with a semicircle attached to one of its longer sides.
What does this symbol most likely represent?
a. A bathtub.
b. A window.
c. A desk.
d. A staircase landing.

Answer: a. A bathtub.
Explanation: This symbol is commonly used to represent bathtubs in blueprints.

401. Case 8: A contractor is examining a blueprint of a commercial building's HVAC system. He notices a symbol that looks like a fan inside a circle.

What does this symbol most likely indicate?
a. Ceiling fan.
b. Exhaust fan.
c. HVAC unit.
d. Ventilation duct.

Answer: b. Exhaust fan.
Explanation: In HVAC blueprints, a fan inside a circle typically represents an exhaust fan.

402. Case 9: On a blueprint, there's a symbol that looks like a T inside a circle. This symbol is located in various rooms throughout the blueprint.
What does this symbol most likely represent?
a. Thermostat.
b. Telephone jack.
c. Transformer.
d. Table location.

Answer: a. Thermostat.
Explanation: The symbol of a T inside a circle typically represents a thermostat on blueprints.

403. Case 10: A contractor is reviewing a blueprint for a new office building. He notices a symbol that looks like a zigzag line inside a rectangle.
What does this symbol most likely represent?
a. Elevator.
b. Staircase.
c. Electrical panel.
d. Radiator.

Answer: d. Radiator.
Explanation: The zigzag line inside a rectangle is a common symbol used to represent radiators in blueprints.

404. In a set of architectural blueprints, a symbol resembling a door with an arc is shown. What does this typically indicate?
a. The location of a window.
b. The swing direction of a door.
c. The height of the door.
d. The material of the door.

Answer: b. The swing direction of a door.
Explanation: The arc attached to the door symbol on blueprints typically indicates the direction in which the door swings open.

405. On a blueprint, you notice a scale notation that reads "1/4" = 1'0"." What does this mean?
a. Every quarter inch on the blueprint represents one foot in real life.
b. Every inch on the blueprint represents a quarter of a foot in real life.
c. Every four inches on the blueprint represents one foot in real life.
d. Every inch on the blueprint represents four feet in real life.

Answer: a. Every quarter inch on the blueprint represents one foot in real life.
Explanation: The scale "1/4" = 1'0"" means that for every quarter inch you measure on the blueprint, it represents one foot in the actual construction.

406. A contractor is reviewing a blueprint and notices a symbol that looks like three parallel horizontal lines getting progressively shorter. What does this symbol most likely represent?
a. Electrical outlet.
b. Staircase.
c. Window.
d. Elevator.

Answer: c. Window.
Explanation: The symbol of three parallel horizontal lines that get progressively shorter is commonly used to represent a window on blueprints.

407. On a blueprint, what does a circle with a number inside, typically located next to specific details or sections, represent?
a. Electrical outlets.
b. Room numbers.
c. Keynotes or general notes.
d. Diameter of a pipe.

Answer: c. Keynotes or general notes.
Explanation: Circles with numbers inside, often found next to specific details or sections on blueprints, are used to reference keynotes or general notes that provide additional information about that detail.

408. In a set of blueprints, a contractor notices a symbol that looks like a small rectangle with a semicircle on top. What does this symbol most likely indicate?
a. A table.
b. A bathtub.
c. A door.
d. A light fixture.

Answer: b. A bathtub.
Explanation: The symbol of a small rectangle with a semicircle on top is commonly used to represent bathtubs in blueprints.

409. On a blueprint, you come across a notation that reads "N.T.S." What does this abbreviation typically stand for?
a. Not to scale.
b. Note this section.
c. Near the stairs.
d. North top side.

Answer: a. Not to scale.
Explanation: "N.T.S." is a common abbreviation on blueprints that stands for "Not to Scale," indicating that the drawing or detail is not drawn to a specific scale.

410. A contractor is reviewing a blueprint and notices a symbol that resembles a lightning bolt. What does this symbol most likely represent?
a. Electrical panel.
b. Grounding.
c. High voltage area.
d. Lighting fixture.

Answer: c. High voltage area.
Explanation: A lightning bolt symbol on blueprints typically indicates a high voltage area or component.

411. On a blueprint, what does a dashed line typically represent?
a. Permanent structures.
b. Hidden or obscured features.
c. Property boundaries.
d. Areas to be demolished.

Answer: b. Hidden or obscured features.
Explanation: Dashed lines on blueprints often represent features that are hidden or obscured from view, such as underground utilities or overhead beams.

412. In a set of blueprints, a symbol that looks like a circle with four smaller circles inside is shown. What does this typically indicate?
a. Electrical outlet.
b. Recessed lighting.
c. HVAC vent.
d. Floor drain.

Answer: d. Floor drain.
Explanation: The symbol of a circle with four smaller circles inside typically represents a floor drain on blueprints.

413. On a blueprint, you notice a series of numbers like "2x4" or "2x6." What are these numbers typically referring to?
a. Room dimensions.
b. Door and window sizes.
c. Lumber dimensions.
d. Electrical circuit numbers.

Answer: c. Lumber dimensions.
Explanation: Numbers like "2x4" or "2x6" on blueprints typically refer to the cross-sectional dimensions of lumber.

414. Which type of blueprint provides a bird's-eye view of a building, showing the layout of rooms, walls, and other architectural features?
a. Elevation view
b. Section view
c. Floor plan
d. Detail view

Answer: c. Floor plan
Explanation: A floor plan provides a top-down view of the layout of a building, detailing the arrangement of rooms, walls, doors, and other features.

415. When a contractor wants to see a side view of a building, showcasing the exterior features and height, which blueprint should they refer to?
a. Axonometric view
b. Elevation view
c. Floor plan
d. Section view

Answer: b. Elevation view
Explanation: Elevation views provide a side perspective of a building, illustrating the exterior features, finishes, and overall height.

416. Which blueprint is typically used to display a vertical cut-through of a building or structure, showing the relationships between different floors and spaces?
a. Elevation view
b. Floor plan
c. Section view
d. Isometric view

Answer: c. Section view
Explanation: Section views offer a vertical cross-section of a building, revealing the relationships and details between different floors, walls, and other structural elements.

417. If a contractor is looking to understand intricate details of a specific component, like a custom window or unique door hinge, which type of blueprint would be most helpful?
a. Floor plan
b. Elevation view
c. Section view
d. Detail view

Answer: d. Detail view
Explanation: Detail views zoom in on specific components or areas, providing in-depth information and dimensions for specialized or custom elements.

418. Which type of blueprint offers a three-dimensional representation without perspective, where all lines are parallel to the three major axes?
a. Perspective view
b. Elevation view
c. Isometric view
d. Floor plan

Answer: c. Isometric view
Explanation: Isometric views depict objects in three dimensions without perspective, with all lines parallel to one of the three major axes.

419. When reviewing a blueprint that showcases the design and appearance of a building's exterior from all sides, you are most likely looking at:
a. Floor plans
b. Elevation views
c. Section views
d. Detail views

Answer: b. Elevation views
Explanation: Elevation views display the exterior design and appearance of a building from all sides, including front, rear, and sides.

420. Which type of blueprint would best illustrate the layout and arrangement of electrical outlets, switches, and light fixtures in a room?
a. Electrical plan
b. Elevation view
c. Floor plan
d. Section view

Answer: a. Electrical plan
Explanation: Electrical plans detail the placement and arrangement of electrical components, such as outlets, switches, and fixtures, within a space.

421. In which blueprint would a contractor find information about the plumbing fixtures, pipe routes, and connections in a building?
a. Floor plan
b. Elevation view
c. Plumbing plan
d. Section view

Answer: c. Plumbing plan
Explanation: Plumbing plans provide details about the placement, routes, and connections of plumbing fixtures and pipes within a building.

422. If a contractor wants to understand how a building looks from a specific angle or vantage point, offering a sense of depth and perspective, which blueprint should they refer to?
a. Isometric view
b. Elevation view
c. Perspective view
d. Floor plan

Answer: c. Perspective view
Explanation: Perspective views depict buildings or objects from a specific vantage point, providing a sense of depth and three-dimensionality.

423. Which blueprint typically showcases the arrangement and type of HVAC components, such as duct routes, vents, and equipment locations?
a. Floor plan
b. HVAC plan
c. Elevation view
d. Section view

Answer: b. HVAC plan
Explanation: HVAC plans detail the layout, type, and routing of HVAC components, including ductwork, vents, and equipment locations.

424. On an architectural blueprint, what does a broken line with alternating long and short dashes typically represent?
a. Property boundary
b. Hidden detail
c. Electrical circuit
d. Water line

Answer: b. Hidden detail
Explanation: In architectural blueprints, a broken line with alternating long and short dashes usually indicates a hidden detail or feature not visible from the current view.

425. Which symbol on an electrical blueprint indicates the location of an electrical outlet?
a. A circle with a single line
b. A circle with two parallel lines
c. A triangle
d. A square

Answer: b. A circle with two parallel lines
Explanation: On electrical blueprints, an electrical outlet is typically represented by a circle with two parallel lines inside.

426. In plumbing blueprints, what does a circle with the letter "V" inside represent?
a. Vent
b. Valve
c. Vessel
d. Vacuum

Answer: a. Vent
Explanation: In plumbing blueprints, a circle with the letter "V" inside typically indicates a vent.

427. On an architectural blueprint, which symbol represents a door?
a. A quarter-circle
b. A rectangle
c. A circle with a cross inside
d. A triangle

Answer: a. A quarter-circle
Explanation: Doors in architectural blueprints are usually represented by a quarter-circle that shows the door's swing direction.

428. In electrical blueprints, how is a light switch typically represented?
a. A circle with an "S"
b. A rectangle with a "T"
c. A circle with a line through it
d. A triangle with a "L"

Answer: c. A circle with a line through it
Explanation: Light switches in electrical blueprints are often depicted as circles with a line through them.

429. On a plumbing blueprint, which symbol would indicate a cold water supply?
a. A circle with a "C"
b. A circle with a "W"
c. A circle with a "CW"
d. A circle with a "H"

Answer: c. A circle with a "CW"
Explanation: In plumbing blueprints, a circle with "CW" inside typically represents a cold water supply.

430. In architectural blueprints, what does a continuous thick line usually represent?
a. Electrical line
b. Property boundary
c. Structural wall
d. Plumbing line

Answer: c. Structural wall
Explanation: In architectural blueprints, a continuous thick line is commonly used to denote a structural wall.

431. On an electrical blueprint, which symbol typically represents a ceiling light?
a. A circle with a "C"
b. A circle with a dot in the center
c. A rectangle with a "L"
d. A triangle with a "T"

Answer: b. A circle with a dot in the center
Explanation: Ceiling lights in electrical blueprints are often depicted as circles with a central dot.

432. In plumbing blueprints, how is a floor drain typically represented?
a. A circle with a cross inside
b. A square with a "D"
c. A triangle with an "F"
d. A circle with a "FD"

Answer: a. A circle with a cross inside
Explanation: Floor drains in plumbing blueprints are usually represented by a circle with a cross inside.

433. On an architectural blueprint, which symbol would indicate a window?
a. A rectangle split into smaller rectangles
b. A triangle with a "W"
c. A circle with a "W"
d. A square with a "W"

Answer: a. A rectangle split into smaller rectangles
Explanation: Windows in architectural blueprints are typically represented by rectangles that might be split into smaller rectangles or squares, showing individual panes.

434. When a blueprint states a scale of 1:50, what does this imply?
a. 1 inch on the blueprint represents 50 feet in real life.
b. 1 centimeter on the blueprint represents 50 meters in real life.
c. 1 foot on the blueprint represents 50 inches in real life.
d. 1 meter on the blueprint represents 50 centimeters in real life.

Answer: b. 1 centimeter on the blueprint represents 50 meters in real life.
Explanation: A scale of 1:50 means that for every 1 unit on the drawing, it represents 50 units in real life.

435. Which of the following scales would show the most detail for a room layout?
a. 1:10
b. 1:50
c. 1:100
d. 1:200

Answer: a. 1:10
Explanation: The smaller the number after the colon, the larger the item appears on the paper, thus showing more detail.

436. On a blueprint with a scale of 1:100, if a wall is represented by 2 centimeters, how long is the actual wall?
a. 2 meters
b. 20 meters
c. 200 meters
d. 100 meters

Answer: b. 20 meters. Explanation: With a 1:100 scale, 1 cm on paper represents 100 cm (or 1 meter) in real life. Thus, 2 cm represents 200 cm, which is 20 meters.

437. If a blueprint does not specify a scale, what should a contractor do?
a. Assume a standard scale of 1:100.
b. Use personal judgment to estimate dimensions.
c. Refer to any dimension lines or notes provided.
d. Discard the blueprint as it's unusable.

Answer: c. Refer to any dimension lines or notes provided.
Explanation: If no scale is specified, dimension lines or specific measurements noted on the blueprint can provide clarity.

438. Which type of scale would most likely be used for a detailed kitchen cabinet drawing?
a. 1:5
b. 1:20
c. 1:50
d. 1:100

Answer: a. 1:5
Explanation: A detailed drawing, such as a kitchen cabinet, would require a larger scale like 1:5 to capture all the intricate details.

439. In a floor plan scaled at 1:200, if a room is 4 cm wide on the blueprint, how wide is it in reality?
a. 8 meters
b. 80 meters
c. 800 meters
d. 20 meters

Answer: b. 80 meters. Explanation: At a 1:200 scale, 1 cm on the blueprint represents 200 cm (or 2 meters) in real life. So, 4 cm represents 800 cm, which is 80 meters.

440. What is the primary purpose of using a scale in blueprints?
a. To fit large objects onto a smaller piece of paper.
b. To make the drawing look more professional.
c. To increase the complexity of the drawing.
d. To reduce the accuracy of measurements.

Answer: a. To fit large objects onto a smaller piece of paper. Explanation: Scales allow designers to represent large real-world dimensions and objects on a much smaller, manageable medium like paper.

441. Which of the following is NOT a type of scale commonly used in blueprints?
a. Linear scale
b. Architectural scale
c. Ratio scale
d. Circular scale

Answer: d. Circular scale
Explanation: While linear, architectural, and ratio scales are common in blueprints, a "circular scale" is not a standard term associated with blueprinting.

442. On an architectural scale where ¼ inch equals 1 foot, how long would a 12-foot wall appear on the blueprint?
a. 3 inches
b. 12 inches
c. 48 inches
d. 0.25 inches

Answer: a. 3 inches
Explanation: If ¼ inch represents 1 foot, then 12 feet would be represented by 12 times ¼ inch, which equals 3 inches.

443. If a contractor realizes the scale on a blueprint is incorrect, what's the best course of action?
a. Adjust the scale mentally and proceed.
b. Contact the designer or architect for clarification.
c. Use a standard scale of 1:100.
d. Measure the actual site and adjust accordingly.

Answer: b. Contact the designer or architect for clarification.
Explanation: If there's any doubt about the accuracy or clarity of a blueprint, it's always best to consult with the person who created it to avoid costly mistakes.

444. Which document typically provides detailed information on the materials and workmanship required for a project?
a. Project schedule
b. Blueprint elevation view
c. Construction contract
d. Specifications

Answer: d. Specifications
Explanation: Specifications are written documents that provide detailed information about the requirements for materials, workmanship, and the methods used in the construction.

445. In a construction project, which document outlines the timeline and sequence of activities?
a. Specifications
b. Project schedule
c. Blueprint section view
d. Material list

Answer: b. Project schedule
Explanation: The project schedule provides an overview of the timeline, milestones, and sequence of activities for the construction project.

446. Which document would a contractor refer to for the terms of payment and other legal considerations?
a. Specifications
b. Blueprint
c. Construction contract
d. Project schedule

Answer: c. Construction contract
Explanation: The construction contract outlines the terms and conditions agreed upon between the client and the contractor, including payment terms, responsibilities, and other legal considerations.

447. A door schedule on a blueprint would provide details about:
a. The timeline for installing doors.
b. The type, size, and location of doors.
c. The contractor responsible for door installation.
d. The cost of each door.

Answer: b. The type, size, and location of doors.
Explanation: A door schedule provides a detailed list of doors, including their type, size, material, and location within the project.

448. Which document would you refer to for a detailed breakdown of individual tasks and their durations in a construction project?
a. Specifications
b. Construction contract
c. Blueprint
d. Work breakdown structure (WBS)

Answer: d. Work breakdown structure (WBS)
Explanation: A WBS provides a detailed breakdown of individual tasks, their sequence, and their durations in a project.

449. Which of the following would NOT typically be found in a construction contract?
a. Payment terms
b. Material specifications
c. Project timeline
d. Client's favorite color

Answer: d. Client's favorite color
Explanation: While a construction contract includes essential details about the project, personal preferences unrelated to the project, like the client's favorite color, would not be included.

450. In the event of a discrepancy between the blueprint and specifications, which document usually takes precedence?
a. Blueprint
b. Specifications
c. Construction contract
d. Project schedule

Answer: b. Specifications. Explanation: Typically, in the event of a discrepancy, the specifications take precedence over the drawings or blueprints.

451. Which document would detail the quality and type of paint to be used in a project?
a. Blueprint
b. Project schedule
c. Construction contract
d. Specifications

Answer: d. Specifications
Explanation: Specifications provide detailed information about materials, including their quality and type, to be used in the construction.

452. What is the primary purpose of a construction contract?
a. To provide a detailed drawing of the project.
b. To list the materials needed for the project.
c. To outline the legal agreement between parties involved in the construction.
d. To provide a timeline for the project.

Answer: c. To outline the legal agreement between parties involved in the construction. Explanation: The construction contract serves as the legal agreement detailing the terms, conditions, responsibilities, and other essential aspects between the client and the contractor.

453. Which document would you refer to for a detailed list of quantities, materials, and labor costs for a project?
a. Blueprint
b. Specifications
c. Bill of Quantities (BOQ)
d. Construction contract

Answer: c. Bill of Quantities (BOQ). Explanation: A BOQ provides a detailed list of materials, their quantities, and associated costs, often used for tendering and project cost assessments.

454. Which of the following best describes the primary purpose of a site plan in a blueprint?
a. To detail the interior finishes of a building.
b. To provide a bird's-eye view of the entire project site.
c. To specify the electrical wiring of a structure.
d. To list the materials needed for the project.

Answer: b. To provide a bird's-eye view of the entire project site. Explanation: A site plan offers an overhead, scaled view of a construction site, showing the existing conditions and proposed construction.

455. On a site plan, what does the symbol of a dashed line typically represent?
a. Existing structures
b. Proposed structures
c. Underground utilities
d. Property boundaries

Answer: c. Underground utilities. Explanation: Dashed lines on site plans often indicate underground utilities or features that aren't immediately visible from an overhead view.

456. In the context of a blueprint, what would you refer to for a detailed view of a specific section of a building, often a vertical cutaway?
a. Elevation
b. Site plan
c. Section view
d. Floor plan

Answer: c. Section view
Explanation: A section view provides a vertical cutaway of a building, offering details about construction and material layers.

457. Which document would typically accompany a blueprint to provide detailed written instructions about materials, quality, and installation procedures?
a. Bill of Quantities (BOQ)
b. Project schedule
c. Specifications
d. Construction contract

Answer: c. Specifications
Explanation: Specifications provide detailed written instructions that accompany blueprints, detailing the quality, materials, and installation procedures.

458. On a blueprint, which symbol typically represents an electrical outlet?
a. A small circle with a line through it
b. A rectangle divided into two parts
c. A triangle pointing downwards
d. A circle with two parallel lines inside

Answer: d. A circle with two parallel lines inside
Explanation: The symbol of a circle with two parallel lines inside typically represents an electrical outlet on blueprints.

459. In a blueprint, how is the North direction typically indicated?
a. With a compass rose
b. With the letter 'N'
c. With an arrow pointing upwards
d. With a dashed line

Answer: b. With the letter 'N'. Explanation: While various symbols can represent North, it's commonly indicated with the letter 'N' on blueprints.

460. Which of the following would NOT typically be found on a site plan?
a. Property boundaries
b. Locations of existing trees and landscaping
c. Interior room layouts
d. Driveway and parking layouts

Answer: c. Interior room layouts
Explanation: Site plans focus on exterior elements. Interior room layouts are typically found in floor plans, not site plans.

461. When interpreting a blueprint, what is the first step a contractor should take?
a. Start the construction immediately.
b. Check the scale of the blueprint.
c. Choose the materials.
d. Assign tasks to the team.

Answer: b. Check the scale of the blueprint.
Explanation: Before starting any work, it's crucial to understand the scale of the blueprint to ensure accurate measurements and construction.

462. On a blueprint, which view provides a horizontal, overhead view of a structure or space?
a. Elevation view
b. Section view
c. Floor plan
d. Site plan

Answer: c. Floor plan
Explanation: A floor plan provides a horizontal, overhead view of the layout of each floor of a structure.

463. In the context of blueprints, what does a clouded or bubbled area typically indicate?
a. An area of focus or detail
b. An area with a mistake or error
c. An area that is not part of the construction
d. An area that has been revised or changed

Answer: d. An area that has been revised or changed. Explanation: Clouding or bubbling an area on blueprints is a common way to highlight revisions or changes made after the initial drawings.

464. Which of the following would be considered a direct cost in a construction project estimate?
a. Office rent
b. Project manager's salary
c. Concrete for the foundation
d. Marketing expenses for the company

Answer: c. Concrete for the foundation
Explanation: Direct costs are those that can be directly attributed to a specific project. Concrete for the foundation is a direct material cost for the project.

465. When calculating labor costs for an estimate, which factor is NOT typically included?
a. Hourly wage
b. Overtime
c. Worker's compensation insurance
d. Client entertainment expenses

Answer: d. Client entertainment expenses
Explanation: Client entertainment expenses are not a direct labor cost. The other options are direct costs associated with labor.

466. A contractor wants to ensure a 20% profit margin on a project. If the total estimated costs are $100,000, what should the contractor charge the client?
a. $120,000
b. $125,000
c. $150,000
d. $80,000

Answer: b. $125,000
Explanation: To achieve a 20% profit margin, the contractor would divide the total costs by (1 - 0.20), which equals $125,000.

467. Which of the following is NOT typically included in overhead costs for a construction project?
a. Equipment rental for the project
b. Utilities for the main office
c. Salaries for administrative staff
d. General marketing expenses

Answer: a. Equipment rental for the project
Explanation: Equipment rental for a specific project is a direct cost. Overhead costs are indirect costs that cannot be attributed to a specific project.

468. In the context of construction estimating, what does "markup" typically refer to?
a. The total of all indirect costs
b. The difference between the total project cost and the desired profit
c. The percentage added to the total costs to determine the bid price
d. The amount subtracted from total costs to determine profit

Answer: c. The percentage added to the total costs to determine the bid price
Explanation: Markup is the percentage added to the cost of a project to determine its selling or bid price.

469. Which of the following would be the LEAST likely to fluctuate during the duration of a construction project?
a. Labor rates
b. Material costs
c. Equipment rental rates
d. Overhead costs

Answer: d. Overhead costs
Explanation: Overhead costs, such as office rent or administrative salaries, are generally fixed and less likely to fluctuate compared to direct costs like labor or materials.

470. When estimating the cost of materials, it's essential to consider:
a. The client's budget
b. The project timeline
c. The contractor's profit margin
d. The number of subcontractors

Answer: b. The project timeline
Explanation: Material costs can change over time due to market conditions, so the project timeline can influence the estimated cost.

471. A construction project has estimated direct costs of $500,000 and overhead costs of $100,000. If the contractor wants a 15% profit margin, what should be the bid price?
a. $600,000
b. $690,000
c. $725,000
d. $575,000

Answer: c. $725,000
Explanation: Total costs are $600,000 ($500,000 + $100,000). To achieve a 15% profit margin, divide the total costs by (1 - 0.15), which equals $725,000.

472. Which of the following is NOT a direct cost in construction estimating?
a. Lumber for framing
b. Fuel for machinery on-site
c. Insurance for the company's fleet of vehicles
d. Wages for the on-site construction crew

Answer: c. Insurance for the company's fleet of vehicles
Explanation: Insurance for the company's fleet of vehicles is an indirect or overhead cost, not directly tied to a specific project.

473. In construction estimating, the term "contingency" refers to:
a. The contractor's desired profit
b. An allowance for unexpected costs
c. The total of all overhead costs
d. Costs associated with project delays

Answer: b. An allowance for unexpected costs
Explanation: A contingency is an amount or percentage added to an estimate to cover unforeseen expenses during the project.

474. Which of the following is the FIRST step in preparing a bid for a construction project?
a. Calculating the total cost
b. Reviewing the project specifications
c. Determining the profit margin
d. Submitting the bid to the client

Answer: b. Reviewing the project specifications
Explanation: Before any calculations or decisions on pricing, a contractor must thoroughly review the project specifications to understand the scope and requirements.

475. When preparing a bid, why is it crucial for contractors to attend pre-bid meetings?
a. To negotiate the contract terms
b. To understand the project's scope and ask questions
c. To meet potential competitors
d. To finalize the bid amount

Answer: b. To understand the project's scope and ask questions. Explanation: Pre-bid meetings allow contractors to get a clearer understanding of the project, clarify doubts, and ensure that their bid is accurate.

476. Which of the following is NOT a common pitfall when preparing a bid?
a. Overestimating material costs
b. Ignoring overhead costs
c. Including a contingency amount
d. Misunderstanding project specifications

Answer: a. Overestimating material costs. Explanation: While overestimating can lead to a higher bid and potentially not winning the project, it's not a pitfall in the same way that underestimating or ignoring costs can be, which can lead to financial losses.

477. In the context of bidding, what does "value engineering" refer to?
a. Reducing the project's scope to fit the budget
b. Offering alternative solutions to decrease costs without compromising quality
c. Increasing the project's value by adding features
d. Engineering the bid to be the lowest among competitors

Answer: b. Offering alternative solutions to decrease costs without compromising quality Explanation: Value engineering aims to improve the value of goods or products and services by examining their function.

478. When determining their bid amount, contractors should ensure:
a. Their bid is the lowest to win the project
b. They have the highest profit margin possible
c. Their bid covers all costs and includes a reasonable profit margin
d. Their bid is at least 20% higher than the estimated costs

Answer: c. Their bid covers all costs and includes a reasonable profit margin. Explanation: While being competitive is essential, contractors must ensure they cover all costs and still make a profit.

479. Which of the following is a common mistake contractors make that can result in an unprofitable project?
a. Over-relying on past project costs
b. Consulting with subcontractors during the bidding process
c. Including a contingency in their bid
d. Reviewing the project specifications multiple times

Answer: a. Over-relying on past project costs
Explanation: Each project is unique, and costs can change. Relying too heavily on past project costs can lead to inaccuracies in the bid.

480. In a competitive bid situation, how can a contractor make their bid stand out without lowering their price significantly?
a. By offering faster completion times
b. By ignoring certain project specifications to reduce costs
c. By increasing the project's scope without client approval
d. By submitting the bid after the deadline to ensure it's the last one seen

Answer: a. By offering faster completion times
Explanation: Offering faster completion times or other value-added services can make a bid more attractive without compromising on price.

481. Which of the following should NOT be included in the final bid submission?
a. A detailed breakdown of costs
b. Proof of insurance and bonding
c. Personal financial statements of the company's owners
d. A proposed project timeline

Answer: c. Personal financial statements of the company's owners
Explanation: Personal financial statements are private and not relevant to the project bid. The other options provide clarity and assurance to the client.

482. When considering subcontractors for a project bid, it's essential to:
a. Always choose the cheapest option to keep costs low
b. Have multiple options in case one falls through
c. Ensure they have a good track record and can meet project requirements
d. Only consider those you've worked with in the past

Answer: c. Ensure they have a good track record and can meet project requirements
Explanation: While cost is a factor, the reliability and quality of work of a subcontractor are paramount.

483. After submitting a bid, a contractor realizes they made a significant error in their calculations. What should they do?
a. Ignore it and hope the client doesn't notice
b. Withdraw their bid immediately
c. Inform the client and provide a corrected bid
d. Wait to see if they win the project before addressing the error

Answer: c. Inform the client and provide a corrected bid
Explanation: Transparency and honesty are crucial. If a mistake is identified, it's best to inform the client and provide the correct information.

484. Which software is commonly used in the construction industry for 3D modeling and aids in estimating by visualizing the project before construction begins?
a. Microsoft Excel
b. AutoCAD
c. Revit
d. Trello

Answer: c. Revit
Explanation: Revit is a Building Information Modeling (BIM) software that allows for 3D design and visualization, aiding in accurate estimating.

485. When using drone technology in construction estimating, what is the primary benefit?
a. Reducing labor costs
b. Providing aerial views for marketing purposes
c. Surveying large or difficult-to-reach areas quickly
d. Replacing the need for on-site visits

Answer: c. Surveying large or difficult-to-reach areas quickly
Explanation: Drones can quickly capture aerial data, making it easier to survey large or challenging sites, which aids in more accurate estimating.

486. Which software is NOT typically used for construction cost estimating?
a. Procore
b. Bluebeam Revu
c. Adobe Photoshop
d. CostX

Answer: c. Adobe Photoshop
Explanation: Adobe Photoshop is primarily a graphic design tool and is not typically used for construction cost estimating.

487. Building Information Modeling (BIM) has become a staple in modern construction estimating. What is its primary advantage?
a. Reducing the need for physical labor
b. Providing a platform for graphic design
c. Allowing for real-time collaboration between teams
d. Replacing traditional 2D blueprints

Answer: c. Allowing for real-time collaboration between teams
Explanation: BIM not only visualizes the project in 3D but also allows for real-time collaboration, ensuring all teams are on the same page, reducing errors and miscommunications.

488. Which of the following is a cloud-based construction management software that aids in estimating by centralizing project data and streamlining workflows?
a. MATLAB
b. Procore
c. SketchUp
d. MS Project

Answer: b. Procore
Explanation: Procore is a cloud-based construction management software that centralizes project data, aiding in more efficient and accurate estimating.

489. Prefabrication is a growing trend in construction. How does it impact the estimating process?
a. Increases the unpredictability of costs
b. Reduces the overall project timeline and associated labor costs
c. Requires more on-site labor
d. Makes the estimating process redundant

Answer: b. Reduces the overall project timeline and associated labor costs
Explanation: Prefabrication means parts are constructed off-site and then assembled on-site, which can reduce the project timeline and associated costs.

490. Which tool is essential for estimators to measure distances and areas on digital construction drawings?
a. Digital calipers
b. On-screen takeoff software
c. 3D printers
d. Virtual reality headsets

Answer: b. On-screen takeoff software
Explanation: On-screen takeoff software allows estimators to measure distances, areas, and volumes directly on digital plans, aiding in accurate estimating.

491. In the context of construction estimating, what does the term "parametric estimating" refer to?
a. Estimating based on unit costs
b. Estimating based on historical data and project parameters
c. Estimating using 3D models
d. Estimating based solely on material costs

Answer: b. Estimating based on historical data and project parameters
Explanation: Parametric estimating uses statistical modeling to determine project costs based on historical data and various project parameters.

492. Which of the following is NOT a benefit of using construction estimating software?
a. Reducing the chance of human error
b. Automatically updating when market prices change
c. Allowing for faster bid preparation
d. Guaranteeing project profitability

Answer: d. Guaranteeing project profitability
Explanation: While estimating software can enhance accuracy and efficiency, it doesn't guarantee profitability. Profitability depends on various factors, including execution and unforeseen project challenges.

493. Virtual Reality (VR) is becoming more prevalent in construction. How can it aid in the estimating process?
a. By replacing the need for physical construction
b. Allowing clients to walk through a project before it's built, potentially leading to early design changes
c. Calculating material costs automatically
d. Serving as the primary tool for drafting blueprints

Answer: b. Allowing clients to walk through a project before it's built, potentially leading to early design changes. Explanation: VR can visualize the project in a way that clients can "experience" it before construction, potentially leading to design changes that can impact the estimate.

494. Which of the following is NOT a common risk associated with project estimating?
a. Fluctuating material prices
b. Labor shortages
c. Overestimation of equipment availability
d. Fixed contract pricing for all projects

Answer: d. Fixed contract pricing for all projects. Explanation: Fixed contract pricing is a method of pricing, not a risk in estimating. The other options represent uncertainties that can affect the accuracy of an estimate.

495. When considering unforeseen site conditions, which strategy can contractors employ during the bidding phase to mitigate potential financial setbacks?
a. Ignoring the conditions until they become problematic
b. Allocating a contingency budget
c. Reducing labor costs to compensate
d. Using cheaper materials

Answer: b. Allocating a contingency budget. Explanation: A contingency budget is set aside to cover unforeseen expenses, ensuring that unexpected site conditions don't lead to financial losses.

496. Which of the following is a risk management technique that involves setting aside a percentage of the project's total estimated cost to cover unforeseen expenses?
a. Value engineering
b. Contingency planning
c. Escalation clause
d. Fixed-price contract

Answer: b. Contingency planning
Explanation: Contingency planning involves setting aside funds (often a percentage of the total estimated cost) to cover unexpected costs during a project.

497. In the context of risk management in estimating, what does an "escalation clause" in a contract refer to?
a. A clause that allows for project termination if costs exceed estimates
b. A clause that outlines penalties for project delays
c. A clause that allows for price adjustments due to unforeseen increases in material or labor costs
d. A clause that mandates fixed pricing for all project elements

Answer: c. A clause that allows for price adjustments due to unforeseen increases in material or labor costs
Explanation: An escalation clause provides a mechanism to adjust prices if certain specified costs (like materials or labor) increase beyond what was originally estimated.

498. Which of the following is NOT a recommended strategy for managing risks associated with labor costs during estimating?
a. Using historical labor cost data
b. Assuming minimum wage for all labor roles
c. Consulting with subcontractors for specialized labor estimates
d. Considering potential overtime or premium time

Answer: b. Assuming minimum wage for all labor roles
Explanation: Assuming minimum wage for all labor roles can lead to significant underestimations. It's essential to consider the actual costs and expertise required for each role.

499. When estimating a project in an area prone to natural disasters, what should a contractor specifically account for to manage potential risks?
a. Reduced material costs
b. Faster project timelines
c. Additional insurance or protective measures
d. Lower labor costs

Answer: c. Additional insurance or protective measures
Explanation: In areas prone to natural disasters, contractors might need to invest in additional insurance or protective measures to safeguard the project, affecting the estimate.

500. Which of the following best describes "value engineering" in the context of risk management in estimating?
a. Reducing project scope to fit the budget
b. Increasing the project budget to fit all desired features
c. A systematic method to improve the "value" of goods or products by examining their function
d. Using the cheapest available materials

Answer: c. A systematic method to improve the "value" of goods or products by examining their function
Explanation: Value engineering aims to increase the value of a project by either improving its function or reducing its cost, ensuring that the project remains within budget without compromising quality.

501. In risk management for estimating, why is it crucial to maintain open communication with suppliers?
a. To negotiate fixed prices for all materials
b. To ensure timely delivery of materials
c. To understand potential fluctuations in material costs
d. To reduce the overall project timeline

Answer: c. To understand potential fluctuations in material costs
Explanation: Open communication with suppliers helps contractors stay informed about potential price changes, ensuring that estimates remain accurate.

502. Which of the following is a potential pitfall contractors should avoid when preparing an estimate?
a. Relying solely on past project data without considering current market conditions
b. Consulting with specialists for areas outside their expertise
c. Including a contingency budget
d. Reviewing the estimate multiple times before submission

Answer: a. Relying solely on past project data without considering current market conditions
Explanation: While past project data is valuable, relying solely on it without considering current market conditions can lead to inaccurate estimates.

503. When considering equipment costs in an estimate, which risk management strategy can help account for potential equipment breakdowns or malfunctions?
a. Ignoring potential equipment issues
b. Allocating funds for equipment maintenance and potential rentals
c. Reducing labor costs to compensate for equipment costs
d. Using only brand-new equipment for every project

Answer: b. Allocating funds for equipment maintenance and potential rentals
Explanation: Setting aside funds for maintenance or potential equipment rentals ensures that work can continue even if equipment breaks down, preventing financial setbacks.

504. Which of the following is a primary responsibility of a construction project manager during the initiation phase of a project?
a. Conducting soil tests
b. Finalizing the project closeout report
c. Defining the project scope
d. Managing subcontractor payments

Answer: c. Defining the project scope
Explanation: During the initiation phase, the project manager's primary responsibility is to define the project scope, which sets the foundation for all subsequent project activities.

505. When managing a construction project, which document is crucial for tracking project progress against planned timelines?
a. Purchase order
b. Gantt chart
c. Change order
d. Warranty certificate

Answer: b. Gantt chart
Explanation: A Gantt chart is a visual representation of a project schedule, showing the start and finish dates of various project elements, making it essential for tracking progress.

506. In the context of construction project management, what is the primary purpose of a "change order"?
a. To initiate a new project
b. To document any changes to the original project scope, timeline, or budget
c. To finalize project deliverables
d. To order new construction materials

Answer: b. To document any changes to the original project scope, timeline, or budget
Explanation: A change order is used to capture any deviations from the original project plan, ensuring all stakeholders are aligned on changes and their implications.

507. Which of the following is NOT typically a responsibility of a construction project manager during the project closeout phase?
a. Securing final permits
b. Conducting a post-project review
c. Initiating subcontractor contracts
d. Ensuring all project deliverables are met

Answer: c. Initiating subcontractor contracts
Explanation: The initiation of subcontractor contracts typically occurs during the planning or execution phases, not during project closeout.

508. When managing risks in a construction project, which strategy involves accepting the consequences of the risk should it occur?
a. Risk avoidance
b. Risk transfer
c. Risk mitigation
d. Risk acceptance

Answer: d. Risk acceptance
Explanation: Risk acceptance means acknowledging the risk and being prepared to deal with the consequences if it materializes, without taking proactive steps to prevent it.

509. Which of the following is a key component of effective stakeholder communication during a construction project?
a. Providing updates only when there are significant project changes
b. Keeping stakeholders informed of project progress and any issues regularly
c. Avoiding sharing bad news to maintain stakeholder confidence
d. Limiting communication to only high-level stakeholders

Answer: b. Keeping stakeholders informed of project progress and any issues regularly
Explanation: Regular and transparent communication ensures that stakeholders are aligned, informed, and can provide necessary feedback or support.

510. In construction project management, what is the primary purpose of "work breakdown structure (WBS)"?
a. To provide a detailed financial analysis of the project
b. To break down the project into smaller, manageable tasks or phases
c. To define the roles and responsibilities of each project team member
d. To outline the project's communication plan

Answer: b. To break down the project into smaller, manageable tasks or phases
Explanation: A WBS decomposes a project into individual tasks or deliverables, helping in planning, assigning responsibilities, and tracking progress.

511. Which of the following is NOT a typical challenge faced by construction project managers?
a. Maintaining project quality
b. Ensuring timely project delivery
c. Deciding the architectural design of the project
d. Managing project costs

Answer: c. Deciding the architectural design of the project
Explanation: While project managers work closely with architects, the decision on architectural design typically rests with the architects and stakeholders, not the project manager.

512. In the context of construction project management, which tool or technique is primarily used for quality assurance and control?
a. SWOT analysis
b. Gantt chart
c. Pareto chart
d. Work breakdown structure

Answer: c. Pareto chart
Explanation: A Pareto chart is a type of chart that contains both bars and a line graph, where individual values are represented in descending order by bars, and the cumulative total is represented by the line. It's used to prioritize the most significant factors in quality control.

513. When managing a construction project, which of the following is crucial for ensuring that the project stays within its budget?
a. Regularly reviewing and updating the project's cost baseline
b. Ignoring minor cost overruns to maintain project momentum
c. Only considering labor costs in the budget
d. Avoiding stakeholder communication to reduce overhead costs

Answer: a. Regularly reviewing and updating the project's cost baseline
Explanation: Regularly reviewing the cost baseline ensures that any deviations from the budget are identified and addressed promptly, ensuring financial control.

514. Which scheduling method visualizes project tasks and their dependencies on a timeline, making it easier to see the entire scope of a project and the relationships between tasks?
a. Critical Path Method (CPM)
b. Gantt chart
c. PERT chart
d. Kanban board

Answer: b. Gantt chart
Explanation: A Gantt chart is a visual representation of a project schedule, showing tasks on a timeline, making it easier to understand the project's flow.

515. In the context of construction scheduling, what does "float" refer to?
a. The amount of time a task can be delayed without delaying the project
b. The initial phase of a project where tasks are not yet defined
c. The time it takes to acquire necessary permits
d. The time allocated for project completion

Answer: a. The amount of time a task can be delayed without delaying the project
Explanation: Float, or slack, refers to the amount of time a task can be postponed without affecting the subsequent tasks or overall project completion date.

516. Which quality control method involves inspecting a random sample of a batch instead of the entire batch?
a. Total Quality Management (TQM)
b. Six Sigma
c. Statistical Process Control (SPC)
d. Quality Assurance (QA)

Answer: c. Statistical Process Control (SPC)
Explanation: SPC uses statistical methods to monitor and control a process to ensure that it operates at its full potential.

517. A construction project is behind schedule due to unforeseen weather conditions. What is the most appropriate immediate action?
a. Ignore the delay and hope to catch up later
b. Inform all stakeholders and adjust the schedule accordingly
c. Cancel the project
d. Allocate all resources to another project

Answer: b. Inform all stakeholders and adjust the schedule accordingly
Explanation: Transparent communication is crucial in project management. Informing stakeholders allows for collaborative problem-solving and sets realistic expectations.

518. Which of the following is NOT a common quality control tool used in construction?
a. Fishbone diagram
b. Pareto chart
c. SWOT analysis
d. Control chart

Answer: c. SWOT analysis
Explanation: While SWOT analysis is a strategic planning tool, it's not typically used for quality control in construction.

519. A subcontractor consistently delivers work that does not meet the project's quality standards. What is the best initial approach to address this?
a. Terminate the subcontractor immediately
b. Communicate the issues and provide an opportunity for correction
c. Ignore the issues and fix them without informing the subcontractor
d. Deduct payment without any communication

Answer: b. Communicate the issues and provide an opportunity for correction
Explanation: Effective communication is key. Addressing issues directly and providing an opportunity for correction promotes collaboration and problem-solving.

520. In construction, "resource leveling" refers to:
a. Ensuring all tasks are of equal importance
b. Adjusting the project schedule to address resource constraints
c. Making sure all floors of a building are at the same height
d. Keeping all resources at the construction site level

Answer: b. Adjusting the project schedule to address resource constraints
Explanation: Resource leveling is a technique in project management that involves adjusting the schedule to ensure that resource usage is kept below certain predefined limits.

521. Which of the following is a proactive approach to quality control in construction?
a. Inspecting work after it's completed
b. Addressing issues only when they become major problems
c. Implementing a quality management system from the project's start
d. Relying solely on client feedback for quality assessment

Answer: c. Implementing a quality management system from the project's start
Explanation: Proactively implementing a quality management system ensures that quality standards are defined and adhered to throughout the project's duration.

522. A construction project is over budget due to unexpected increases in material costs. What should be the immediate step?
a. Continue the project without any changes
b. Inform stakeholders and explore cost-saving measures
c. Reduce the quality of materials to save costs
d. Stop the project indefinitely

Answer: b. Inform stakeholders and explore cost-saving measures
Explanation: Keeping stakeholders informed allows for collaborative decision-making, and exploring cost-saving measures can help bring the project back within budget.

523. In the context of construction, which of the following best describes "lead time"?
a. The time it takes for a leader to make a decision
b. The time between starting a task and its completion
c. The time between ordering a material and its delivery
d. The time allocated for breaks during construction

Answer: c. The time between ordering a material and its delivery
Explanation: Lead time refers to the duration between the order of a material or service and its delivery or completion. It's crucial for scheduling and ensuring materials arrive when needed.

524. Which OSHA standard mandates that employers provide a workplace free from recognized hazards?
a. OSHA 1910.120
b. OSHA 1910.38
c. OSHA 1910.95
d. OSHA 5(a)(1)

Answer: d. OSHA 5(a)(1)
Explanation: OSHA 5(a)(1) is also known as the General Duty Clause, which requires employers to provide a workplace free from recognized hazards that could cause death or serious harm.

525. Fall protection is required at elevations of _____ or more in general industry.
a. 4 feet
b. 6 feet
c. 10 feet
d. 12 feet

Answer: a. 4 feet
Explanation: In general industry, fall protection is required at elevations of 4 feet or more.

526. Which of the following is NOT a requirement for a ladder used in construction?
a. Side rails must extend at least 3 feet above the landing
b. Ladders must be free of sharp edges and splinters
c. Ladders can be coated with a slippery material to enhance their appearance
d. Ladders must be inspected regularly

Answer: c. Ladders can be coated with a slippery material to enhance their appearance
Explanation: OSHA standards prohibit the use of slippery material on ladders as it can pose a safety risk.

527. What is the primary purpose of OSHA's Hazard Communication Standard (HCS)?
a. To ensure that employers provide free PPE
b. To ensure that information about chemical hazards is communicated to employees
c. To mandate the use of specific chemicals in the workplace
d. To regulate the storage of all chemicals on a job site

Answer: b. To ensure that information about chemical hazards is communicated to employees
Explanation: OSHA's HCS is designed to ensure that information about chemical and toxic substance hazards in the workplace and associated protective measures is disseminated to workers.

528. Which OSHA standard addresses the requirements for fire extinguishers in the workplace?
a. OSHA 1910.1200
b. OSHA 1910.157
c. OSHA 1910.134
d. OSHA 1910.146

Answer: b. OSHA 1910.157
Explanation: OSHA 1910.157 specifically covers the placement, use, maintenance, and testing of portable fire extinguishers provided for the use of employees.

529. In which situation would an employer be exempt from maintaining an OSHA 300 Log?
a. The company employs more than 10 people.
b. The company is in a high-risk industry.
c. The company employs only family members.
d. The company employs 10 or fewer employees during the year.

Answer: d. The company employs 10 or fewer employees during the year.
Explanation: Small businesses with 10 or fewer employees are exempt from many of OSHA's recordkeeping requirements, including maintaining an OSHA 300 Log.

530 What is the maximum allowable noise exposure level for an 8-hour workday, according to OSHA?
a. 85 dBA
b. 90 dBA
c. 95 dBA
d. 100 dBA

Answer: b. 90 dBA
Explanation: OSHA's permissible exposure limit (PEL) for noise exposure in general industry is 90 dBA for an 8-hour workday.

531. Which of the following is NOT a requirement for safety training and education under OSHA standards?
a. Training in a language the worker understands
b. Training only once at the time of hire
c. Periodic retraining when new hazards are introduced
d. Training on the specific hazards of the job

Answer: b. Training only once at the time of hire
Explanation: OSHA requires training to be ongoing, especially when new hazards are introduced or when the worker is assigned to a new area with different hazards.

532. Under OSHA standards, when must a hard hat be worn?
a. Only during inspections
b. When there's a potential for head injury from falling objects
c. Only during the initial construction phase
d. At all times on a job site

Answer: b. When there's a potential for head injury from falling objects
Explanation: OSHA requires hard hats to be worn when there's a potential for head injury from falling objects, electrical shocks, or other impacts.

533. Which of the following is a primary focus of OSHA's confined spaces standard?
a. Ensuring workers have enough space to work comfortably
b. Regulating the temperature inside confined spaces
c. Ensuring workers can enter and exit safely
d. Mandating a minimum of two exits for all confined spaces

Answer: c. Ensuring workers can enter and exit safely
Explanation: OSHA's confined spaces standard primarily focuses on ensuring that workers can safely enter and exit confined spaces, and that they are protected from hazards inside these spaces.

534. Which OSHA standard specifically addresses safety requirements for electrical installations?
a. OSHA 1926.400
b. OSHA 1910.120
c. OSHA 1926.1053
d. OSHA 1910.95

Answer: a. OSHA 1926.400
Explanation: OSHA 1926.400 specifically addresses electrical safety requirements that are necessary for the practical safeguarding of employees involved in construction work.

535. For plumbers working in trenches, OSHA requires a protective system for trenches that are how deep?
a. 3 feet or deeper
b. 4 feet or deeper
c. 5 feet or deeper
d. 6 feet or deeper

Answer: c. 5 feet or deeper
Explanation: OSHA requires a protective system for trenches that are 5 feet or deeper unless made entirely in stable rock.

536. Masonry workers often deal with silica. What is OSHA's permissible exposure limit (PEL) for respirable crystalline silica for construction?
a. 25 µg/m^3 over an 8-hour TWA
b. 50 µg/m^3 over an 8-hour TWA
c. 75 µg/m^3 over an 8-hour TWA
d. 100 µg/m^3 over an 8-hour TWA

Answer: b. 50 µg/m^3 over an 8-hour TWA
Explanation: OSHA has set the permissible exposure limit for respirable crystalline silica in construction at 50 micrograms per cubic meter of air, averaged over an 8-hour shift.

537. Which OSHA standard is most relevant to carpenters working on scaffolding?
a. OSHA 1926.451
b. OSHA 1910.134
c. OSHA 1926.1052
d. OSHA 1910.147

Answer: a. OSHA 1926.451
Explanation: OSHA 1926.451 provides general requirements for scaffold safety, ensuring that carpenters and other workers are protected when working on scaffolds.

538. Electrical workers must maintain a safe distance from overhead power lines. What is the minimum distance that must be maintained from a 50kV overhead line?
a. 5 feet
b. 10 feet
c. 15 feet
d. 20 feet

Answer: d. 20 feet
Explanation: For voltages over 50kV, workers must maintain a distance of 20 feet plus 0.4 inches for each 1kV over 50kV.

539. Which of the following is NOT a requirement for masonry saws?
a. Must be equipped with a guard
b. Must be used dry, without water
c. Must be equipped with a tight-fitting and well-maintained shroud
d. Must be used with a commercially available dust collection system

Answer: b. Must be used dry, without water
Explanation: Masonry saws should be used with water delivery system to suppress dust when dry cutting.

540. For plumbers, what is the primary focus of OSHA's standard on "Medical Services and First Aid"?
a. Ensuring access to a first aid kit
b. Ensuring access to a hospital within 15 minutes of a worksite
c. Ensuring plumbers are trained in CPR
d. Ensuring plumbers have received tetanus shots

Answer: a. Ensuring access to a first aid kit
Explanation: OSHA's standard on "Medical Services and First Aid" primarily requires employers to ensure the ready availability of medical personnel for advice and consultation and emphasizes the importance of quick aid for injured employees.

541. In carpentry, when using pneumatic nail guns, what safety feature is OSHA particularly concerned with?
a. Sequential triggers
b. Compressor pressure gauges
c. Nail size selectors
d. Adjustable exhausts

Answer: a. Sequential triggers
Explanation: Sequential triggers reduce the risk of accidental nail discharge, which can be a significant hazard in carpentry.

542. For electrical workers, what is the primary protective equipment when working on live circuits?
a. Safety goggles
b. Rubber gloves
c. Steel-toed boots
d. Earplugs

Answer: b. Rubber gloves
Explanation: Rubber gloves provide insulation from electrical currents, making them essential protective equipment for electrical workers dealing with live circuits.

543. Masonry workers are often at risk of which of the following due to the materials they work with?
a. Electrocution
b. Silicosis
c. Hearing loss
d. Radiation exposure

Answer: b. Silicosis
Explanation: Masonry workers often deal with materials that produce silica dust when cut or ground. Inhaling this dust can lead to silicosis, a lung disease.

544. Which of the following is a mandatory piece of safety equipment for construction workers when there is a potential for falling objects?
a. Safety goggles
b. Steel-toed boots
c. Hard hats
d. Earplugs

Answer: c. Hard hats
Explanation: Hard hats are essential for protecting workers from potential head injuries due to falling objects.

545. When working in a trench that is 5 feet deep or more, which safety equipment is mandated by OSHA?
a. Safety harness
b. Protective system (e.g., shoring, shielding)
c. Steel-toed boots
d. Respirator

Answer: b. Protective system (e.g., shoring, shielding)
Explanation: OSHA requires a protective system for trenches that are 5 feet or deeper to prevent cave-ins.

546. Which of the following is NOT a mandatory piece of safety equipment stipulated by OSHA for construction workers exposed to high noise levels?
a. Earplugs
b. Safety goggles
c. Earmuffs
d. Hard hats

Answer: b. Safety goggles
Explanation: While safety goggles are essential for eye protection, they are not specifically for noise protection. Earplugs and earmuffs are used to protect against high noise levels.

547. When working at heights above 6 feet, OSHA mandates the use of:
a. Safety nets
b. Fall protection systems
c. Steel-toed boots
d. Safety goggles

Answer: b. Fall protection systems
Explanation: OSHA requires the use of fall protection systems when workers are operating at heights of 6 feet or more above a lower level.

548. For a construction worker involved in welding operations, which safety equipment is crucial to prevent flash burns?
a. Hard hats
b. Welding shields or goggles
c. Insulated gloves
d. Steel-toed boots

Answer: b. Welding shields or goggles
Explanation: Welding shields or goggles protect the eyes from the intense light and UV radiation produced during welding, preventing flash burns.

549. Which of the following incidents does NOT need to be reported to OSHA within 24 hours?
a. An amputation
b. A work-related in-patient hospitalization of three employees
c. A work-related injury requiring first aid
d. Loss of an eye

Answer: c. A work-related injury requiring first aid
Explanation: Injuries requiring only first aid are not subject to the 24-hour reporting requirement. However, amputations, in-patient hospitalizations, and loss of an eye must be reported within that time frame.

550. If a contractor is found in repeated violation of OSHA standards, they may face:
a. Mandatory retraining
b. A written warning
c. Increased inspection frequency
d. Significantly higher penalties

Answer: d. Significantly higher penalties
Explanation: Repeated violations can result in significantly higher penalties, showcasing OSHA's commitment to ensuring workplace safety.

551. Which of the following is a primary responsibility of employers regarding Personal Protective Equipment (PPE) under OSHA standards?
a. Ensure PPE is personally purchased by employees
b. Provide training on PPE once a year
c. Ensure PPE is maintained in a sanitary and reliable condition
d. Ensure PPE is of a universal size

Answer: c. Ensure PPE is maintained in a sanitary and reliable condition
Explanation: OSHA mandates that employers ensure the PPE is maintained in a sanitary and reliable condition to protect employees from hazards.

552. In the event of a severe incident, OSHA's investigation primarily aims to:
a. Assign blame to the workers involved
b. Determine the cause and ensure similar incidents are prevented
c. Penalize the contractor irrespective of the cause
d. Ensure insurance claims are valid

Answer: b. Determine the cause and ensure similar incidents are prevented
Explanation: OSHA's primary goal is to promote safety. Investigations aim to determine the cause of incidents and ensure measures are in place to prevent similar future occurrences.

553. Which of the following is NOT a factor OSHA considers when determining the penalty for a violation?
a. The company's profit margins
b. The gravity of the violation
c. The size of the business
d. The employer's history of violations

Answer: a. The company's profit margins
Explanation: While OSHA considers the gravity of the violation, the size of the business, and the history of violations, the company's profit margins are not a direct factor in penalty determination.

554. If a contractor disagrees with an OSHA citation, they:
a. Have no right to appeal
b. Must immediately stop all ongoing projects
c. Can formally contest the citation
d. Are mandated to pay double the penalty

Answer: c. Can formally contest the citation
Explanation: Contractors have the right to formally contest an OSHA citation if they disagree with it.

555. Which of the following best describes OSHA's primary goal?
a. Increase federal revenue through penalties
b. Ensure worker safety and health
c. Reduce the number of businesses in operation
d. Promote the use of PPE sales

Answer: b. Ensure worker safety and health
Explanation: OSHA's primary mission is to ensure safe and healthful working conditions for working men and women by setting and enforcing standards and by providing training, outreach, education, and assistance.

556. Which of the following is the primary reason for maintaining a clean and organized job site?
a. To impress clients
b. To ensure efficient workflow
c. To prevent potential hazards and injuries
d. To save on cleaning costs at the end of a project

Answer: c. To prevent potential hazards and injuries
Explanation: While all the options have their merits, the primary reason for maintaining a clean and organized job site is to prevent potential hazards and ensure the safety of all workers.

557. When operating a power tool, it's essential to:
a. Use the tool as quickly as possible to save time
b. Remove safety guards for better access
c. Wear appropriate PPE and follow manufacturer's instructions
d. Use the tool for multiple purposes to save costs

Answer: c. Wear appropriate PPE and follow manufacturer's instructions
Explanation: Safety is paramount. Using tools according to the manufacturer's instructions and wearing the appropriate PPE can prevent accidents and injuries.

558. A worker on a construction site is required to work at a height of 10 feet. Which of the following PPE is mandatory according to OSHA standards?
a. Safety goggles
b. Fall protection system
c. Steel-toed boots
d. Earplugs

Answer: b. Fall protection system
Explanation: OSHA requires the use of fall protection systems when workers are operating at heights of 6 feet or more above a lower level.

559. Which of the following is NOT a common hazard associated with electrical tools?
a. Electrocution
b. Burns
c. Slips and falls due to cords
d. Loud noise exposure

Answer: d. Loud noise exposure
Explanation: While electrical tools can pose several risks, loud noise exposure is typically associated with tools like jackhammers or saws, not specifically electrical hazards.

560. In the event of a small fire at a job site, the FIRST action should be to:
a. Attempt to put it out with a fire extinguisher
b. Evacuate the area and alert others
c. Take a photo for documentation
d. Call the project manager

Answer: b. Evacuate the area and alert others
Explanation: Safety first. Before attempting any other actions, it's crucial to ensure everyone is safe and aware of the danger.

561. Which of the following is NOT a purpose of a Safety Data Sheet (SDS) on a construction site?
a. Provide information on potential hazards of a product
b. Offer first-aid measures in case of exposure
c. Detail the company's financial data
d. Describe protective measures and PPE recommendations

Answer: c. Detail the company's financial data
Explanation: SDSs are meant to provide safety and health-related information about products, not financial data of a company.

562. When using a ladder, it's essential to:
a. Use the top step for added height
b. Ensure the ladder is on stable and level ground
c. Lean as far as possible without moving the ladder
d. Use a metal ladder near electrical sources for durability

Answer: b. Ensure the ladder is on stable and level ground
Explanation: For safety, always ensure the ladder is stable and on level ground to prevent falls and accidents.

563. Which of the following is a primary reason for conducting regular safety meetings on a job site?
a. To take a break from work
b. To fulfill insurance requirements
c. To ensure all workers are updated on safety protocols and potential hazards
d. To socialize with coworkers

Answer: c. To ensure all workers are updated on safety protocols and potential hazards
Explanation: Regular safety meetings ensure that all workers are aware of and can discuss safety protocols, potential hazards, and any changes in the work environment.

564. A worker is exposed to loud noise from machinery on a job site. Which PPE is most appropriate to protect against potential hearing damage?
a. Safety goggles
b. Hard hat
c. Earplugs or earmuffs
d. Respirator

Answer: c. Earplugs or earmuffs
Explanation: Earplugs or earmuffs are designed to protect against high noise levels and prevent potential hearing damage.

565. In the event of an accident on a job site, what is the FIRST step a contractor should take?
a. Document the incident for insurance
b. Ensure the injured party receives appropriate medical attention
c. Call the project owner
d. Review safety protocols to prevent future incidents

Answer: b. Ensure the injured party receives appropriate medical attention
Explanation: The immediate well-being of the injured individual is the top priority. Ensuring they receive medical attention is the first step before addressing other concerns.

566. A worker has suffered a minor cut on his arm. What is the FIRST step in treating the wound?
a. Apply a tourniquet above the wound.
b. Clean the wound with soap and water.
c. Cover the wound with a bandage.
d. Apply an antibiotic ointment.

Answer: b. Clean the wound with soap and water.
Explanation: The first step in treating a minor cut is to clean it to prevent infection. Using soap and water is the most basic and effective method.

567. If a worker is experiencing a heat stroke, which of the following actions should you NOT take?
a. Move the person to a cooler place.
b. Give the person a drink of water.
c. Apply ice directly to the person's skin.
d. Fan the person.

Answer: c. Apply ice directly to the person's skin.
Explanation: Applying ice directly can cause cold injury. Instead, use cool cloths or a lukewarm shower.

568. A colleague has inhaled a harmful chemical fume. What is the immediate course of action?
a. Make them lie down and rest.
b. Move them to fresh air immediately.
c. Give them water to drink.
d. Induce vomiting.

Answer: b. Move them to fresh air immediately.
Explanation: If someone has inhaled harmful fumes, the priority is to get them to an area with fresh air to prevent further inhalation of the toxic substance.

569. Which of the following is NOT a sign of a broken bone?
a. Swelling or bruising over a bone.
b. Intense thirst.
c. Deformity of an arm or leg.
d. Pain in the injured area.

Answer: b. Intense thirst.
Explanation: Thirst is not a direct symptom of a broken bone. The other options are common signs of a fracture.

570. In case of an electrical burn, what should be your primary concern?
a. Applying cold water to the burn.
b. Checking the person's heart and breathing.
c. Covering the burn with a cloth.
d. Removing the person from the electrical source.

Answer: d. Removing the person from the electrical source.
Explanation: The first step is to ensure the person is no longer in contact with the electrical source to prevent further injury.

571. If a worker is bitten by a snake, which of the following should NOT be done?
a. Keep the person calm.
b. Apply a tourniquet above the bite.
c. Immobilize the bitten area.
d. Seek medical attention immediately.

Answer: b. Apply a tourniquet above the bite.
Explanation: Applying a tourniquet can trap the venom in one area, making it more concentrated and potentially more harmful.

572. A colleague has a foreign object embedded in their eye. What should you do?
a. Try to remove the object with tweezers.
b. Rub the eye to dislodge the object.
c. Rinse the eye with water.
d. Cover the eye with a bandage and seek medical attention.

Answer: d. Cover the eye with a bandage and seek medical attention.
Explanation: Do not try to remove the object or rinse the eye, as this can cause further damage. Covering the eye prevents further injury, and immediate medical attention is required.

573. Which of the following is a symptom of shock?
a. Flushed skin.
b. Thirst.
c. Rapid breathing.
d. Over-excitement.

Answer: c. Rapid breathing.
Explanation: Rapid breathing, along with clammy skin, rapid heartbeat, and dizziness, are common symptoms of shock.

574. A worker has suffered a chemical burn. After ensuring safety, what is the next step?
a. Apply a neutralizing agent.
b. Rinse the burn with plenty of water.
c. Dry the area with a cloth.
d. Apply a thick layer of burn ointment.

Answer: b. Rinse the burn with plenty of water.
Explanation: Rinsing the burn with water helps to remove the chemical and reduce the severity of the burn.

575. If someone is suspected of having a spinal injury, what should you NOT do?
a. Keep the person still.
b. Hold the person's head to prevent movement.
c. Move the person unless there's immediate danger.
d. Offer the person a drink.

Answer: c. Move the person unless there's immediate danger.
Explanation: Moving someone with a suspected spinal injury can cause further harm. It's essential to keep them still unless there's an immediate threat, like a fire.

576. A contractor is considering entering into a joint venture with another firm for a specific project. Which of the following is NOT typically a characteristic of a joint venture?
a. Shared profits and losses.
b. A separate legal entity from the parent companies.
c. Limited to a specific project or duration.
d. Joint management and control.

Answer: b. A separate legal entity from the parent companies.
Explanation: A joint venture is typically not a separate legal entity but rather a contractual arrangement between parties for a specific project.

577. Which of the following best describes a "mechanic's lien"?
a. A claim made by a subcontractor for unpaid labor.
b. A claim against a property by someone who supplied labor or materials for work on that property.
c. A legal claim against a mechanic for faulty work.
d. A warranty provided by construction equipment manufacturers.

Answer: b. A claim against a property by someone who supplied labor or materials for work on that property.
Explanation: A mechanic's lien is a legal claim against a property by a contractor, subcontractor, or supplier who hasn't been paid.

578. In a cost-plus contract, the contractor is reimbursed for:
a. Only the direct costs of labor and materials.
b. Direct costs plus a fixed fee.
c. Direct costs plus a percentage of the total project cost.
d. Only the indirect costs associated with overhead.

Answer: c. Direct costs plus a percentage of the total project cost.
Explanation: In a cost-plus contract, the contractor is reimbursed for the direct costs of a project and also receives a fee, typically a percentage of the total project cost.

579. Which of the following is NOT typically covered under a general liability insurance policy for contractors?
a. Injuries to employees on the job site.
b. Damage to client property caused by the contractor's operations.
c. Claims of false advertising.
d. Bodily injury claims from third parties.

Answer: a. Injuries to employees on the job site.
Explanation: Injuries to employees are typically covered under workers' compensation insurance, not general liability insurance.

580. A contractor's license has been suspended. What is the most appropriate next step for the contractor?
a. Continue working but only on small projects.
b. Appeal the suspension if there are grounds to do so.
c. Transfer the license to a colleague.
d. Ignore the suspension and continue business as usual.

Answer: b. Appeal the suspension if there are grounds to do so.
Explanation: If a contractor believes the suspension is unjust, the appropriate action is to appeal. Working without a valid license can lead to further penalties.

581. Which of the following best describes "indemnification" in a construction contract?
a. A clause that determines how disputes will be resolved.
b. A provision that requires one party to compensate another for certain losses.
c. A guarantee that the project will be completed on time.
d. A clause specifying payment terms.

Answer: b. A provision that requires one party to compensate another for certain losses.
Explanation: Indemnification is a contractual obligation of one party to compensate the loss occurred to the other party due to the act of the indemnitor or any other party.

582. In the context of contract law, what does the term "consideration" refer to?
a. Deliberation and discussion before signing a contract.
b. The thoughtfulness of each party toward the other.
c. Something of value exchanged between parties.
d. A review of contract terms after a dispute arises.

Answer: c. Something of value exchanged between parties.
Explanation: Consideration refers to something of value (like money, a service, or a promise) that is exchanged between parties in a contract.

583. Which of the following is NOT a primary purpose of a business plan?
a. To establish the company's mission and objectives.
b. To guide the company's operations and decision-making.
c. To secure financing or investment.
d. To serve as a daily checklist for operations.

Answer: d. To serve as a daily checklist for operations. Explanation: While a business plan provides guidance and direction, it is not typically used as a daily operational checklist.

584. A contractor is found to be in breach of contract. What potential remedies are available to the injured party?
a. Specific performance.
b. Termination of the contract.
c. Damages.
d. All of the above.

Answer: d. All of the above.
Explanation: The injured party may seek various remedies depending on the nature of the breach, including specific performance, termination, or damages.

585. Which of the following best describes "liquidated damages" in a construction contract?
a. Damages paid to liquidate or end the contract.
b. A predetermined amount of money that must be paid as damages for breach of contract.
c. The total sum of all damages incurred during the project.
d. A refund provided to the client for unsatisfactory work.

Answer: b. A predetermined amount of money that must be paid as damages for breach of contract.
Explanation: Liquidated damages are a predetermined amount agreed upon by both parties at the time of contract formation that will be paid if a specific breach occurs.

586. In the critical path method (CPM) of scheduling, what does the term "float" refer to?
a. The amount of time a task can be delayed without delaying the project.
b. The total duration of the project.
c. The earliest start time of a task.
d. The tasks that are not critical to project completion.

Answer: a. The amount of time a task can be delayed without delaying the project.
Explanation: In CPM, float (or slack) is the amount of time a task can be delayed without causing a delay to subsequent tasks or the project completion date.

587. Which of the following is NOT a primary function of a building envelope?
a. Structural support.
b. Aesthetic appeal.
c. Temperature regulation.
d. Soundproofing.

Answer: d. Soundproofing.
Explanation: While a building envelope can contribute to soundproofing, its primary functions are to provide structural support, protect against external elements, and regulate the internal environment.

588. In a reinforced concrete beam, the purpose of adding steel reinforcement is to:
a. Increase its compressive strength.
b. Increase its tensile strength.
c. Make it lighter.
d. Prevent concrete spalling.

Answer: b. Increase its tensile strength.
Explanation: Concrete is strong in compression but weak in tension. Steel reinforcement is added to increase the tensile strength of the beam.

589. Which foundation type is particularly suitable for soils with low bearing capacity, like expansive clays or loose sands?
a. Slab-on-grade.
b. Strip foundation.
c. Pile foundation.
d. Raft foundation.

Answer: c. Pile foundation.
Explanation: Pile foundations transfer loads to deeper, more stable soil layers or even to bedrock, making them suitable for soils with low bearing capacity.

590. The concept of "building orientation" primarily affects which of the following?
a. Structural integrity.
b. Aesthetic appeal.
c. Energy efficiency and comfort.
d. Cost of construction.

Answer: c. Energy efficiency and comfort.
Explanation: Building orientation, especially in relation to the sun, plays a significant role in passive solar design, affecting the building's energy efficiency and internal comfort.

591. In masonry construction, what is the primary purpose of a "weep hole"?
a. To allow for air circulation.
b. To provide an aesthetic detail.
c. To allow moisture to escape from the wall cavity.
d. To reduce the weight of the wall.

Answer: c. To allow moisture to escape from the wall cavity.
Explanation: Weep holes are small openings that allow any accumulated moisture inside the wall cavity to drain out, preventing water damage and related problems.

592. Which of the following is NOT a key consideration when selecting construction materials for a project?
a. Color and aesthetic appeal.
b. Cost and availability.
c. Popularity and trendiness.
d. Durability and maintenance requirements.

Answer: c. Popularity and trendiness.
Explanation: While trends can influence design choices, the primary considerations for material selection are functionality, cost, aesthetics, durability, and maintenance.

593. In green or sustainable construction, what is the primary purpose of a "green roof"?
a. To provide additional recreational space.
b. To reduce the heat island effect and manage stormwater.
c. To add aesthetic appeal to the building.
d. To increase the structural load on the building.

Answer: b. To reduce the heat island effect and manage stormwater.
Explanation: Green roofs, covered with vegetation, help in reducing the heat island effect by cooling the surrounding air and also manage stormwater runoff.

594. Which construction method is particularly known for its speed of erection and is commonly used for skyscrapers?
a. Masonry construction.
b. Timber framing.
c. Steel frame construction.
d. Tilt-up construction.

Answer: c. Steel frame construction.
Explanation: Steel frame construction allows for prefabricated components, making the erection process faster. It's commonly used for high-rise buildings.

595. In terms of indoor air quality, which of the following is a primary concern in modern tightly sealed buildings?
a. Inadequate structural ventilation.
b. Too much natural light.
c. Excessive noise from the outside.
d. Thermal bridging.

Answer: a. Inadequate structural ventilation.
Explanation: Tightly sealed buildings can trap pollutants indoors, making adequate ventilation crucial for maintaining good indoor air quality.

596. In electrical trade, what color is typically used to identify a ground wire in a standard electrical cable?
a. Red
b. Black
c. White
d. Green or bare

Answer: d. Green or bare
Explanation: In standard electrical wiring, the ground wire is typically either green or bare without insulation.

597. For plumbing, what type of pipe is best suited for potable water supply due to its resistance to corrosion and scaling?
a. PVC
b. Copper
c. Galvanized steel
d. ABS

Answer: b. Copper
Explanation: Copper pipes are commonly used for potable water supply lines because of their corrosion resistance and long life.

598. In masonry, which type of mortar is known for its high compressive strength and is commonly used in foundation walls and other load-bearing applications?
a. Type N
b. Type O
c. Type M
d. Type S

Answer: c. Type M
Explanation: Type M mortar has the highest compressive strength, making it suitable for foundations and other structural applications.

599. For carpentry, which type of joint is often used to join two pieces of wood at right angles, such as in the corners of a frame?
a. Dovetail joint
b. Mortise and tenon joint
c. Miter joint
d. Butt joint

Answer: c. Miter joint
Explanation: A miter joint is made by beveling the two pieces to be joined, usually at a 45-degree angle, to form a 90-degree corner.

600. In HVAC, which refrigerant is known for being environmentally friendly and is increasingly used as a replacement for older, ozone-depleting refrigerants?
a. R-22
b. R-410A
c. R-12
d. R-134a

Answer: b. R-410A
Explanation: R-410A is a hydrofluorocarbon (HFC) which does not contribute to ozone depletion, making it a more environmentally friendly choice.

601. For roofing, which type of material is known for its longevity and fire-resistant properties?
a. Asphalt shingles
b. Wood shakes
c. Slate tiles
d. Rubber membrane

Answer: c. Slate tiles
Explanation: Slate tiles can last for decades, even centuries, and are naturally fire-resistant.

602. In the painting trade, which type of paint finish is best suited for high-traffic areas due to its durability and ease of cleaning?
a. Matte
b. Satin
c. Semi-gloss
d. Gloss

Answer: c. Semi-gloss
Explanation: Semi-gloss paints are durable, resist moisture, and can be cleaned easily, making them suitable for high-traffic areas.

603. For tile setting, which adhesive is commonly used to set tiles in wet areas like bathrooms?
a. Mastic
b. Thinset mortar
c. Epoxy adhesive
d. Construction adhesive

Answer: b. Thinset mortar
Explanation: Thinset mortar is water-resistant and provides a strong bond, making it ideal for wet areas.

604. In landscaping, which type of plant is specifically grown and used to prevent soil erosion on slopes?
a. Annuals
b. Groundcovers
c. Shrubs
d. Trees

Answer: b. Groundcovers
Explanation: Groundcovers spread across the soil surface, helping to hold the soil in place and prevent erosion.

605. For drywall installation, what is the primary purpose of "joint compound"?
a. To adhere drywall panels to studs.
b. To fill and smooth seams between drywall panels.
c. To provide insulation behind the drywall.
d. To add a decorative texture to the drywall surface.

Answer: b. To fill and smooth seams between drywall panels.
Explanation: Joint compound, often referred to as "mud", is used to fill the seams between drywall panels and to cover fasteners, creating a smooth surface.

In Closing... As we draw the curtains on this comprehensive journey through the vast world of contracting, I want to pause for a moment and reflect on the ground we've covered together. From the intricate details of blueprints to the nuances of OSHA standards, from the art of estimating to the science of job site safety, we've delved deep into the heart of what it means to be a contractor in today's dynamic world.

Remember, the world of construction isn't just about bricks, beams, and bolts. It's about dreams - the dreams of those who envision magnificent structures and the dreams of those like you, who turn these visions into reality. Along the way, there will be challenges, unforeseen obstacles, and moments of self-doubt. But with every challenge comes an opportunity to learn, grow, and redefine your limits.

It's natural to fear the unknown, to second-guess decisions, or to feel overwhelmed by the sheer scale of a project. But let these moments not deter you. Instead, let them be the stepping stones that propel you to greater heights. Embrace your failures, not as setbacks, but as lessons that sharpen your skills and deepen your understanding.

In this vast arena, you're not alone. There's a community of professionals, mentors, and peers, all of whom have faced similar challenges and emerged stronger. Lean on them, learn from them, and remember to offer a helping hand to those who come after you.

As you step out, armed with knowledge and bolstered by determination, remember that success in this field isn't just about technical expertise. It's about integrity, dedication, and the unwavering belief in one's ability to bring dreams to life.

So, as you turn the page to the next chapter of your professional journey, I want to leave you with this: Believe in yourself, trust the process, and know that every stone you lay, every beam you set, and every space you create is a testament to your passion, skill, and the dreams you're helping to realize.

Best of luck, dear builder of dreams. The world awaits your creations.

Made in the USA
Las Vegas, NV
14 March 2024

87187089R10138